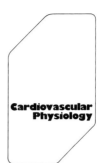

Cardiovascular Physiology

Karger Continuing Education Series

Henry S. Badeer
Omaha, Nebr.

Cardiovascular Physiology

A Synopsis

188 figures and 4 tables, 1984

Basel · München · Paris · London · New York · Tokyo · Sydney

Henry S. Badeer, M.D., Professor, Department of Physiology, Creighton University School of Medicine, 2500 California Street, Omaha, NE 68178

Karger Continuing Education Series, Vol. 6

Topics covered in the Karger Continuing Education Series are selected to help improve clinical skills and introduce the reader to health-related areas undergoing exceptional growth. Produced as compact instructive texts, volumes set forth information which serves to heighten the general awareness and command of current medical procedures and practice. The concise textbook format enhances the value of these books as convenient teaching and training tools for medical scientists, medical clinicians, and health professionals.

National Library of Medicine, Cataloging in Publication
Badeer, Henry S.
Cardiovascular physiology: a synopsis
Henry S. Badeer.—Basel; New York: Karger, 1984
(Karger continuing education series; v. 6)
1. Cardiovascular system—physiology I. Title II. Series
W1 KA821P v. 6
[WG 102 B133c]
ISBN 3-8055-3796-4

Drug Dosage
The author and the publisher have exerted every effort to ensure that drug selection and dosage set forth in this text are in accord with current recommendations and practice at the time of publication. However, in view of ongoing research, changes in government regulations, and the constant flow of information relating to drug therapy and drug reactions, the reader is urged to check the package insert for each drug for any change in indications and dosage and for added warnings and precautions. This is particularly important when the recommended agent is a new and/or infrequently employed drug.

Contents

The Heart

The Blood Vessels

Cardiovascular Regulation

Special Circulations

Appendix

Preface

There is a need for a concise yet sufficiently advanced textbook on cardiovascular physiology to meet the requirements of medical students whose knowledge in this field must exceed that of other students in the health sciences. This book is essentially based on the lectures and handouts that I have given to medical students over many years of teaching. The subject is tersely presented in a somewhat outline form with brief explanations adequate for an average student. I consider excessive explanations as being too much ''spoon feeding'' which, in my opinion, leads to an undesirable passivity, especially in a subject like cardiovascular physiology. Further justification for the terse presentation is the limited teaching time alloted to the medical physiology course. With the explosion of information in all basic medical sciences and the time restrictions in medical curricula, it is imperative to be brief and to the point. This objective, I believe, has been achieved.

The small type used in some parts of the text (eg, equations) is in accordance with the format of the book and does not imply that the material is of less importance than the rest.

The extensive use of illustrations as a teaching tool is an important aspect of the book. Some figures are original and many have been redrawn for the purpose.

The reader will note that many applied topics such as the physiology of cardiac arrhythmias, circulatory shock, hypertension, heart failure, and so on, have not been covered. Likewise, the circulation of the skin, splanchnic organs, kidneys, uterus, and so forth are not included. No comprehensive review of the cardiovascular changes in muscular exercise is presented. However, the effects of exercise on many aspects of circulation are discussed in different sections of the book and the student should, at the end of the course, be able to integrate the overall changes in circulation during muscular exercise. This will provide the student a good opportunity to test his or her ability to put pieces together in an orderly manner. After all, the practice of medicine requires a great deal of analysis and synthesis.

It gives me great pleasure to thank Mrs. Maureen Billones for her meticulous care in typing the manuscript. Thanks are also due Mr. Antranig Chelebian, my technician at the American University of Beirut, for the many diagrams he has drawn in the past. My gratitude to the many authors and publishers who have given permission to use their illustrations or to redraw them for this text.

I take the sole responsibility for any errors bound to occur and would welcome comments and criticism from students and colleagues alike.

Henry S. Badeer

This book is dedicated to medical students whose unbiased inquiry often leads to better understanding

Introduction

Evolution of the Circulatory System

In unicellular and small multicellular organisms, the diffusion and active membrane transport of materials into and out of cells are adequate to meet their metabolic needs. In large multicellular organisms the cells are too numerous and too distant from the ambient medium in which the organism lives. Thus, the direct transfer of materials between cells and the *external* ambient cannot occur. Hence, a system of *mass transport* (bulk flow) of a fluid medium has evolved. This function is carried out by the circulatory system, which consists of a pump (the heart) and a system of vessels. Some of the vessels (arteries) distribute the blood; others serve to exchange materials with cells or the ambient medium (the capillaries) while still others (veins) collect and return the blood to the pump.

The complexity of the circulatory system varies not only with the total number of cells but also with the metabolic rate of the cells. The latter is affected markedly by the temperature of the cells, which is kept at a high quite constant level (37°–40°C) in "warm-blooded" animals (birds and mammals).

The Heart

1 Anatomic Considerations

The adult human heart, weighing about 300g, consists of two pumps fused together into one organ. The advantages of this arrangement will be considered under the subject of cardiac output in Chapter 9. Each pump has its own vascular circuit, and the two circuits are placed "in series" with each other—the left pump feeding into the right and the right into the left (fig. 1).

The *left ventricle* or pump drives the blood into the systemic circuit which is designed for the delivery of nutrients and substrates for the function of *all the cells* of the body and for the removal of waste products and, with few exceptions, removal of heat derived from cellular activity. In addition, this circuit takes up essential materials absorbed from the digestive tract and also delivers blood to the organs of excretion, notably the kidneys. No less important is the uptake of hormones secreted by various endocrine organs to be carried to the target organs.

The right ventricle pumps blood into the pulmonary circuit primarily for gas exchange, but the pulmonary circulation and the lungs have other secondary functions too (blood reservoir function, activation of angiotensin I to angiotensin II, and so on). Note that the lung parenchyma receives most of its nutrients from the systemic circuit by way of the bronchial arteries, which arise from the thoracic aorta (fig. 1).

The pumping of blood by the ventricles is made possible by the presence of valves, which, depending on the anatomic orientation, close or open to permit unidirectional flow. The anatomic details of cardiac valves should be reviewed by the reader at this time. Each ventricle has a receiving chamber (the atrium), which plays a relatively minor role in pumping blood. It serves mainly as a temporary storage depot for blood returning from the veins during the period of contraction and the early part of relaxation of the ventricles when blood cannot enter into the ventricles (atrioventricular valves being closed). Consequently, the atrial walls are thin. Patients in whom atria do not beat properly (eg,

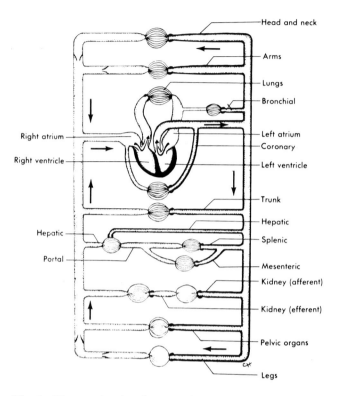

Head and neck

Arms

Lungs

Bronchial

Right atrium

Left atrium

Coronary

Right ventricle

Left ventricle

Trunk

Hepatic

Hepatic

Splenic

Portal

Mesenteric

Kidney (afferent)

Kidney (efferent)

Pelvic organs

Legs

Fig. 1. Diagram showing the general arrangement of systemic and pulmonary circuits placed in series with one another. Note the numerous parallel vascular beds that characterize the systemic circuit. [Reproduced with permission from Berne, R.M.; Levy, M.N.: Cardiovascular Physiology, 4th ed. (C.V. Mosby, St. Louis 1981).]

atrial fibrillation) cannot engage in heavy exercise but can carry on ordinary activities of life. Pumping of blood into each circuit is done by the ventricles, which have thicker walls than the atria. In adult individuals the left ventricular wall is about three times as thick as the right because the systemic circuit has a much *higher resistance* to blood flow than the pulmonary circuit and therefore a *higher pressure gradient* is required to maintain the *same flow* through the two circuits.

Since the pressure developed by the two ventricles is markedly different, the anatomic design of the ventricles is strikingly dissimilar. The left ventricle is described as a *prolate ellipse* (or ellipsoidal) with thick walls to develop a high pressure (fig. 2). The orientation of the fibers in the wall of the left ventricle is complex, as shown in figure 3. The

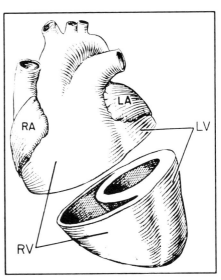

Fig. 2. Diagram showing the shape and relationship of the two ventricles to each other. The left is ellipsoidal and the right is crescent shaped, being wrapped around the left ventricle. [Reproduced with permission from Guyton, A.C.: Textbook of Medical Physiology, 5th ed. (W.B. Saunders, Philadelphia, 1976).]

Fig. 3. Arrangement of muscle fibers from the middle of the free wall of the left ventricle from a heart in systole. The sections are parallel to the epicardial surface and are arranged sequentially. [Reproduced, with permission of the American Heart Association, from Streeter D.D., Jr.; Spotnitz, H.M.; Patel, D.P.; Ross, J. Jr.; Sonnenblick, E.H.: Fiber orientation in the canine left ventricle during diastole and systole. Circ. Res. 24: 339 (1969).]

ENDOCARDIUM

MID-
WALL

EPICARDIUM

septum is functionally more a part of the left ventricle than that of the right. It is curved in both the transverse and the longitudinal sections. The right ventricle is *semilunar* in shape overlying the ventricular septum and has thin walls that are related to the low pressure developed in the cavity under normal circumstances (fig. 2). Many diagrams in texts misrepresent these anatomical features which are adaptive in nature.

Heart muscle may be divided into two major categories: the ordinary (or typical) and the specialized (or atypical) muscle fibers. The ordinary fibers are designed to contract, develop tension and shorten, thereby propelling blood. The specialized fibers initiate and conduct impulses. However, the typical fibers can also initiate impulses under certain conditions and the specialized fibers can contract to a slight extent since they have some myofibrils.

Heart muscle cells (myocytes) are very similar to skeletal muscle cells with regard to striations. The basic unit is the sarcomere extending from one Z line to the next Z line and consisting of thick filaments of myosin (in the A band) and thin filaments of actin (extending from the Z lines through the I band) (fig. 4). However, heart muscle cells (fibers) differ in that they have a single large *centrally* located nucleus. Furthermore, the fibers branch and connect with each other running in different directions. The end of cells makes contact with the end of other cells at the *intercalated discs* which represent the apposition of the sarcolemma at the end of two adjacent cells. At certain points in the disc there are regions of intimate contact called the *nexus* or *gap junction* (fig. 5).

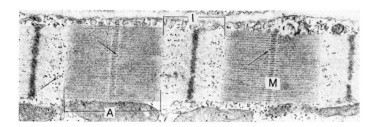

Fig. 4. Cat atrial muscle. Longitudinal section of a fiber showing cross striations. The thick filaments make up the A-band (A) in the center of which is a dense area, the M-band (M). The thin filaments running transversely to the long axis of the fiber are seen in the region of the M-band. The I-band (I) consists of the thin filaments and is bisected by the Z-line. Glycogen granules occur in the cytoplasm and between the myofilaments (arrows). The mitochondria can be seen below. [Reproduced with permission from Katz, A.M.: Physiology of the Heart (Raven Press, New York 1977).]

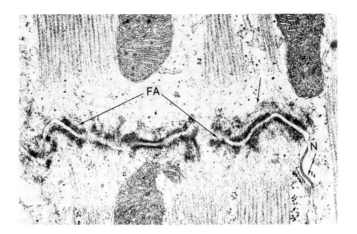

Fig. 5. Cat ventricular muscle showing a transverse section of an intercalated disc. The plasma membranes of adjacent cells are separated by 200–250 Å of extracellular space. The thin actin filaments attach to a dense structure whose filaments run parallel to the cytoplasmic surface of the membrane. These filaments attach to a structure known as the fascia adherens (FA). Near the intercalated disc the plasma membrane comes into intimate contact at a nexus (N) or gap junction, which is believed to be the low resistance pathway between adjacent muscle cells. [Reproduced, with permission, from McNutt, N.S.; Fawcett, D.W.: Myocardial ultrastructure; in Langer and Brady: The Mammalian Myocardium (John Wiley, New York 1974).]

This region is believed to have a very low electrical resistance (400 times lower) compared with the resistance of the sarcolemma along the long axis of the fiber (lateral surface membrane) so that an action potential in one fiber can easily be conducted to all muscle fibers of the heart in all directions. Thus, although the heart consists of millions of discrete anatomically separate cells, it acts *functionally* as one cell, described as a *syncytium* (many cells acting as one cell). For this reason an excitation at any point in the heart travels to all parts of the heart provided other parts are not in an inexcitable refractory state. This is in sharp contrast to skeletal muscle in which each fiber is excited independently of the others by its motor nerve fiber, permitting spatial summation. The property of conduction is known as the dromotropic property.

Another important characteristic is that heart muscle cells are *very rich in mitochondria,* which constitute about 25%–30% of the muscle mass. This correlates well with the high oxygen consumption of the heart and its inability to undergo any significant "oxygen debt" (having very limited anaerobic metabolism). The high energy expenditure of the heart is related to its repetitive mechanical activity associated with the development of tension in the walls at short intervals (75 times/min under resting conditions).

To provide the necessary oxygen and other metabolic fuels, heart muscle fibers have a smaller diameter than skeletal and are richly supplied with blood capillaries. There are invaginations of the sarcolemma that run transversely into the fiber at the region of the Z lines known as the *transverse* or *T-tubules* (fig. 6). These tubules are believed to transmit the action potential from the surface sarcolemma to the interior of the fibers thereby facilitating excitation of the fibers. In addition, there is an extensive network of smaller diameter tubules, running longitudinally, which belong to the system known as the *sarcoplasmic reticulum* (SR). Some parts of the SR are close to the T-system as well as the surface sarcolemma. The functional value of these structural features

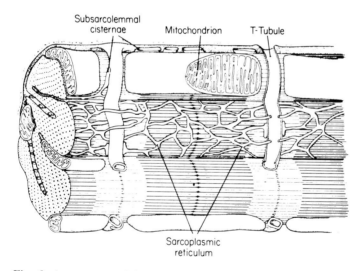

Fig. 6. Arrangement of the sarcoplasmic reticulum in a myocardial fiber. The cell membrane of the muscle fiber has tubular projections (T or transverse tubules) that penetrate into the fiber at right angles to the surface, thereby connecting the extracellular space with the inside of the fiber. Another completely intracellular membrane system, known as the sarcoplasmic reticulum (SR), makes close contact with the T-tubules and the sarcolemma. [Reproduced, with permission, from Smith, J.J.; Kampine, J.P.: Circulatory Physiology—The Essentials. (Williams and Wilkins, Baltimore 1980).]

is to facilitate diffusion between the interstitial fluid of the muscle and the intracellular constituents of the muscle fibers.

The specialized muscle tissue of the heart consists of the S-A node, the internodal tracts between the S-A node and the A-V node, the A-V node, the bundle of His and its two branches and the diffuse Purkinje network of the two ventricles. These tissues are also syncytial and are designed to initiate and conduct impulses.

The two atria are functionally separated from the two ventricles by a ring of *connective tissue*, which forms a framework (fibrous skeleton) to which the cardiac valves and muscles are attached. The fibrous tissue, being inexcitable, cannot conduct impulses from the atria to the ventricles. The only functional connection normally is by way of the A-V node and the bundle of His. In a sense, therefore, the two atria constitute one syncytial mass of muscle and, similarly, the two ventricles, another syncytial mass. This separation gives added importance to the functional connection between the atria and the ventricles by way of the A-V node and the bundle of His. Pathologic processes in this tissue interfere with the normal coordination of atrioventricular activity (refer to the cardiac cycle in Chapter 6).

References

McNutt, N.S.; Fawcett, D.W.: Myocardial ultrastructure; in Langer and Brady: Mammalian Myocardium; pp. 1–49 (John Wiley, New York 1974).

Spiro, D.: The fine structure and contractile mechanism of heart muscle; in Briller and Conn: The Myocardial Cell; pp. 13–61 (University of Pennsylvania Press, Philadelphia 1966).

2 Some Functional Characteristics of Heart Muscle

This section is intended to point out some of the important characteristics (electrical and mechanical) of cardiac muscle tissue. To facilitate the understanding, a few salient features of resting membrane potential of cardiac muscle are briefly reviewed. It is assumed that the student has fully covered, elscwhere in the physiology course, the details of resting membrane and action potentials in skeletal muscle.

Different parts of the heart exhibit characteristic features both as to the magnitude of the resting potentials as well as to the shape and duration of the action potentials (fig. 7). The resting membrane potential of atrial, ventricular and Purkinje fibers is about -90 mV, which refers to the inside of the cell. The origin of this potential is attributable to differences—across the cell membrane—in the concentration of various ions, chiefly Na^+, K^+, Cl^-, HCO_3^- and a large intracellular anion designated as A^- (Ca^{2+} and other ions including HCO_3^- contribute very little to the resting membrane potential). Na^+ and Cl^- are chiefly outside the cell and K^+ and A^- are mostly inside (table 1). The factors that affect this distribution are: (a) differences in the permeability of the membrane to these ions; (b) diffusional forces due to concentration gradients across the membrane; (c) electrostatic forces due to membrane polarity, inside being negative; and (d) existence of active transport mechanisms (pumps) in the membrane, particularly for Na^+ and K^+.

The following facts are recognized to play a key role in causing the membrane potential and the normal distribution of ions across the "resting" cell membrane. The resting membrane is much more permeable to the diffusion of K^+ than to Na^+ (hydrated Na^+ is larger than hydrated K^+ and passes with difficulty through specific channels). The large negatively charged A^- is confined to the inside of the cell and cannot diffuse out through the membrane. The cell membrane contains a coupled Na^+–K^+ pump (active transport using energy derived from ATP) that drives the Na^+ out of the cell and K^+ into the cell. The

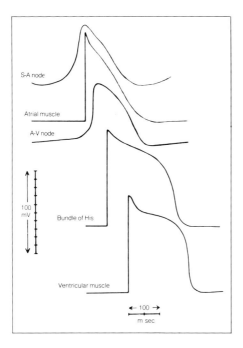

S-A node

Atrial muscle

A-V node

100
mV   Bundle of His

Ventricular muscle

← 100 →
m sec

Fig. 7. Cardiac transmembrane action potentials recorded in vitro from various sites arranged in sequence. [Redrawn from Hoffman, B.F.; Cranefield, P.F.: Electrophysiology of the Heart (McGraw-Hill Book Co., New York 1960).]

number of Na^+ extruded by the pump slightly exceeds the number of K^+ transferred into the cell. Therefore, the pump creates a potential difference across the cell membrane and is described as an *electrogenic pump*. However, it has been estimated that this accounts for a relatively small fraction of the membrane voltage (less than 10 mV). Similarly, A^- inside the cell accounts for only a few millivolts of the -90 mV. This points to the most important factor determining the magnitude of the membrane potential, ie, the role of the differential *permeability* of the membrane to Na^+ and K^+. As the pump drives the Na^+ out of the cell and the K^+ in, it creates a large concentration gradient of these two ions across the cell membrane. Consequently, K^+ diffuses out of the cell through K^+ channels (passive back diffusion or efflux) and Na^+ diffuses in through Na^+ channels (passive back diffusion or influx), but the membrane is much more permeable to K^+ than to Na^+, as already pointed out. Therefore, *K^+ diffuses out much faster than Na^+ diffuses*

Table 1. Approximate electrolyte composition of cardiac intracellular and extracellular fluid in vitro

Ion	Concentration (mmol/L)		Equilibrium potential (mV)[c]
	Intracellular[a]	Extracellular[b]	
Na$^+$	7	144	+81
K$^+$	151	4	−97
Cl$^-$	4[d]	114	−90

[a]Data from Robertson, W. Van B.; Dunihue, F.W.: Water and electrolyte distribution in cardiac muscle. Am J Physiol 177:292(1954).
[b]Data from Pitts, R.F.: The Physiological Basis of Diuretic Therapy (Charles C. Thomas, Springfield, Ill. 1959).
[c]Calculated using the Nernst equation at body temperature.
[d]Calculated from membrane potential.
Reproduced by permission from Little, R.C.: Physiology of the Heart and Circulation; 2nd ed. (Year Book Medical Publishers, Chicago 1981).

in, thereby creating a net accumulation of positive charges outside the cell. Thus, the major part of − 90 mV intracellular potential is caused by the *difference* in membrane permeability to K$^+$ as compared with Na$^+$. Permeability is often expressed as conductance (gNa$^+$, gK$^+$).

One of the major findings that explains the origin of transmembrane potential is the existence of active transport (or pump) in the membrane. This concept utilizes, among others, quantitative methods of physical chemistry in which *concentration gradients* of charged particles (ions) can be related quantitatively to *voltage gradients* across a semipermeable membrane. Concentration and electrical gradients can oppose and neutralize each other to maintain a state of equilibrium. Nernst has expressed this quantitatively by an equation where:

$$E_{ion} = \frac{61}{Z} \log_{10} \frac{[ion]_{out}}{[ion]_{in}}$$

where E$_{ion}$ = millivolts required to maintain a given concentration gradient of an ion across a membrane; Z = valence of the ion and its sign; [ion]$_{out}$ = concentration of the ion outside the membrane; [ion]$_{in}$ = concentration of the same ion inside the membrane. E$_{ion}$ is called the *equilibrium potential* for that ion.

Table 1 shows the equilibrium potential for Na^+, K^+ and Cl^- as determined from concentration gradients across the cardiac cell membrane. The equilibrium potential for Na^+ in heart muscle is about $+80$ mV. This means that to maintain 144 millimoles of Na^+/L outside the cell vs. 7 mmol/L inside, one needs a voltage of $+80$ mV *inside* to oppose it. The equilibrium potential for K^+ is -97 mV for heart muscle. In other words, to maintain 4 mmol K^+/L outside vs. 151 mmol K^+/L inside, one needs to have -97 mV inside the cell to oppose it. The equilibrium potential for Cl^- is -90 mV. This means that the normal resting membrane potential of -90 mV is just the right voltage to oppose and maintain the high concentration of Cl^- outside the cell. Hence, chloride is *passively* distributed across the resting heart cell and requires no pump to maintain the normal concentration gradient.

The K^+ equilibrium potential of about -97 mV is not very far from the normal -90 mV. If K^+ was passively distributed across the cardiac cell membrane, its concentration inside would have been about 114 mmol/L vs. 4 mmol/L outside (from Nernst equation). However, the K^+ inside heart muscle is about 151 mmol/L, which indicates that there is an active transport mechanism (pump) driving into the cell an additional 37 mmol/L of K^+ ($151-114=37$). This extra concentration gradient is equivalent to an electrical force of 7 mV $[-90mV -(-97 mV)=7 mV]$.

The situation is quite different for Na^+. The equilibrium potential for Na^+ is about $+80$ mV inside the cell. But normally the inside of the cell has an opposite potential of -90 mV. Hence, there is an imbalance between the normal electrical and the concentration gradients for Na^+. Thus, passive forces cannot explain Na^+ distribution. Therefore, an active pump is necessary to drive the Na^+ out of the cell. Opposing the pump are the concentration gradient and the voltage gradient both of which tend to drive the Na^+ into the cell. However, the passive influx of Na^+ into the cell is *small* because membrane permeability is low.

Before proceeding further, review the following aspects of action potentials in skeletal muscle: threshold potential, Na^+ and K^+ permeability, depolarization, Hodgkin cycle, overshoot, repolarization, refractory period, temporal and spatial summation of contractions, and tetanus.

Autorhythmicity of the Heart

One of the most striking characteristics of the heart is its ability to excite itself without the necessity of extrinsic nerves (isolated perfused heart beats for many hours or days; transplanted heart beats for years). This property is referred to as *autorhythmicity* or chronotropic property.

Autorhythmicity is essentially a property of specialized heart muscle, best developed in the S–A node, A–V node and Purkinje tissue. Microelectrode studies of these tissues demonstrate a spontaneous depolarization during diastole called *pacemaker potential* or *diastolic depolarization* which on reaching a threshold voltage fires an action potential (figs. 8 and 9).

The ionic basis for pacemaker potential is believed to be a time-dependent *slow decrease in the* K^+ *permeability* (or K^+ conductance) *of the cell membrane* ($gK^+\downarrow$). It is sometimes described as a slow inactivation of the outward K^+ current. There is a progressive decrease in the outward diffusion of K^+ from the pacemaker cells. This means that more K^+ remains inside the cell which becomes progressively less negative (or depolarizes) until the threshold potential is reached. At this point, the action potential is triggered and a marked increase in Na^+ permeability and Hodgkin cycle take over. According to some investigators pacemaker potential may be due to a slow inward Ca^{2+} current.

The tissue that has the highest slope (steepness) of pacemaker potential will discharge at the fastest frequency (other things being equal) and dominate and determine the rate of the entire heart. Normally, this is the S-A node, which is described as the *normal pacemaker* (fig. 10). Some of the findings supporting this view are:

1. Cooling and warming the S-A node (but not other areas of the heart) change the frequency of the heart.
2. Surface electrical recording shows electronegativity of the S-A node occurs earlier than in any other area of the heart.
3. True pacemaker cells can be found in S-A tissue, with the microelectrode. (fig. 11).

The frequency of discharge of pacemaker cells can be altered in several ways from the standpoint of membrane potentials (fig. 11):

1. Change in *slope* of pacemaker potential (line *ab*).
2. Shift in *threshold potential* or voltage (point *b* moves up or down).
3. Change in *maximal diastolic potential* (point *a* moves up or down).

We shall note later that changing the slope of pacemaker potential is the most common way of altering the frequency of discharge (activity of the sympathetic nerves to the heart increasing the slope and that of the vagi to the heart, decreasing the slope of pacemaker potential; cooling the pacemaker decreases the slope and slows the heart rate and vice versa).

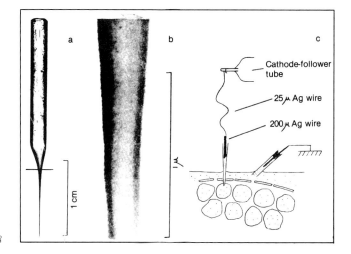

8

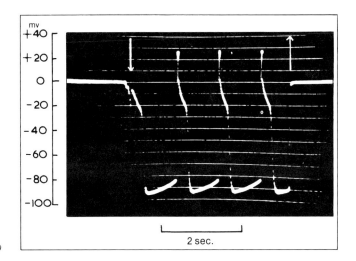

9

Fig. 8. Diagram of the setup for recording the transmembrane potentials of a cardiac muscle fiber. (a) and (b) show the glass microelectrode, (c) represents the "riding" microelectrode impaling a single fiber of a contracting strip of heart muscle. [Reproduced with permission from Weidmann, S.: Resting and action potentials of cardiac muscle. Ann. N.Y. Acad. Sci. *65:* 663 (1957).]

Fig. 9. Isolated Purkinje fiber of a dog heart. A microelectrode was introduced into a single fiber and then withdrawn. Note the spontaneous "pacemaker" potentials that trigger the action potentials. [Reproduced with permission from Weidmann, S.: Resting and action potentials of cardiac muscle. Ann. N.Y. Acad. Sci. *65:* 663 (1957).]

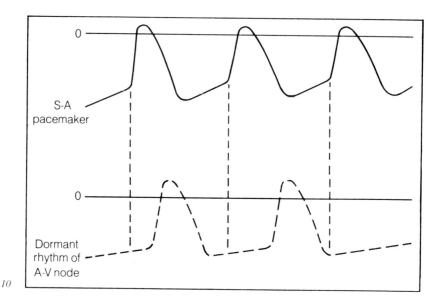

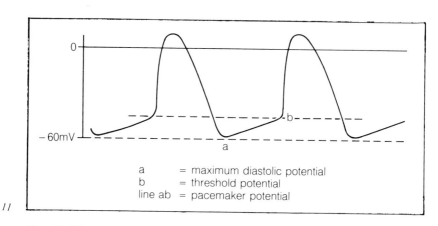

a   = maximum diastolic potential
b   = threshold potential
line ab   = pacemaker potential

Fig. 10. Diagram showing how the faster rhythm of the S-A node overrides the slower rhythm of the A-V node which remains dormant in the normal heart. The passage of impulses arising in the S-A node normally suppresses the rhythmicity of the A-V node and other pacemaker tissues of the heart.

Fig. 11. Diagram illustrating the possible ways in which the pacemaker frequency of the S-A node may be altered: change in level a; change in level b; change in slope of line ab (most common).

Duration of Action Potential

An important characteristic of typical heart muscle is the *long dura-tion* of its action potential. In skeletal muscle, the action potential lasts a few milliseconds whereas in heart muscle it may last 200 msec or more. This is due to the long "plateau" (flat portion) of the action potential, particularly in the A-V node, Purkinje tissue and the typical ventricular muscle tissue (fig. 12). The short rapid repolarization after the upstroke is due to the decrease in fast Na^+ conductance as well as a transient increase in Cl^- conductance causing inward flow of Cl^- (from high concentration outside to low inside). The ionic basis of the plateau seems to involve changes in membrane permeability to the following three ions (fig. 13):

1. Decreased K^+ conductance of membrane ($gK^+\downarrow$), sometimes de-scribed as "anomalous rectification".
2. Activation of a second Na^+ channel, which is much slower then the initial fast Na^+ channel. It is referred to as the slow inward Na^+ current in contrast to the fast Na^+ current responsible for the up-stroke of the action potential.

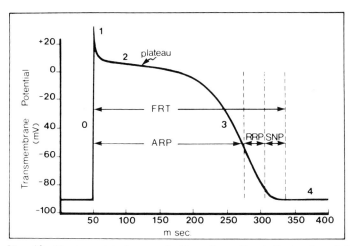

Fig. 12. Single ventricular muscle fiber. Recording indicates the relation between transmembrane action potential and excitability of fiber to cathodal stimulation. ARP = absolute refractory period; RRP = relative refractory period; SNP = supernormal pe-riod; FRT = full recovery time. The various phases of action potential are labeled 0, 1, 2, 3, and 4. [Reproduced with permission from Hoffman, B.F.; Cranefield, P.F.: Elec-trophysiology of the Heart (McGraw-Hill Book Co., New York 1960).]

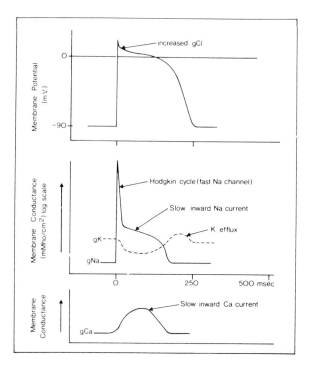

Fig. 13. Changes in membrane permeability to various ions during the action potential of a ventricular muscle fiber (nonrhythmic).

3. Increased membrane conductance to Ca^{2+} causing a *slow* inward Ca^{2+} current ($Ca^{2+}\uparrow$). This Ca^{2+} seems to be responsible for triggering the release of Ca^{2+} from the sarcotubular system by increasing its membrane permeability.

During the final phase of repolarization, K^+ conductance is believed to *increase above* the resting value (heart cells lose small amounts of K^+), although some claim that it simply returns to its high resting value. At this time Na^+ and Ca^{2+} conductances have already returned to their resting values. The membrane of the sarcotubular system pumps Ca^{2+} back into the tubules by an active process. All these processes repolarize the cell to the resting level. During the action potential, the small amount of K^+ that has leaked out and the Na^+ that has entered the cell are replaced by the $Na^+–K^+$ pump during the subsequent rest period. Presumably the Ca^{2+} that has entered through the membrane is also removed from the cells.

inexcitable or *refractory* during this period. From the beginning of the action potential to about the middle of the final repolarization the muscle does not respond to a second stimulus, no matter how strong. This is the *absolute refractory period* (ARP) (fig. 12). The ARP is followed by a *relative* refractory period (RRP) during which a suprathreshold stimulus can evoke a propagated response or action potential. In some experimental studies there is a phase of hyperexcitability after RRP, called *the supernormal phase*, during which a stimulus *below* resting threshold can cause a response. The physiologic significance of this phase is unclear. It should be emphasized that refractoriness is a *membrane property* and is not related to the mechanical activity of the muscle, namely, the interaction between actin and myosin triggered by Ca^{2+}.

The long refractory period of heart muscle prevents the heart from being tetanized by repeated stimulation. Tetanus in skeletal muscle is a sustained contraction without being interrupted by relaxations. To function as a pump the heart must relax and refill. Tetanic contraction would be catastrophic to the organism! Thus, the long refractory period is a functional adaptation for the survival of the organism. On the other hand, in skeletal muscle summation of contractions and tetanus are exceedingly important in permitting graded contractions which are so valuable for controlling its performance, which should vary according to functional requirements. Here, a *short refractory* period is an adaptation for the function of skeletal muscle.

The duration of contraction of heart muscle is relatively long compared with the single twitch of skeletal muscle. Some evidence suggests that it may be related, at least partly, to the long duration of the action potential. The functional value of such duration would be to allow time to pump adequate volumes of blood *per beat,* specially in *large-sized* mammalian hearts where the heart rate is slow. Also, the relatively slow speed of contraction tends to avoid excessive aortic peak (systolic) pressure which is partly determined by the speed of ventricular ejection of blood (other things being equal).

Excitation-contraction Coupling

As in skeletal muscle, the action potential in heart muscle is followed by mechanical contraction after a short interval (5–10 msec) (fig. 14). This relationship is referred to as the excitation-contraction coupling or *E-C coupling*. During this period Ca^{2+} enters the myocardial

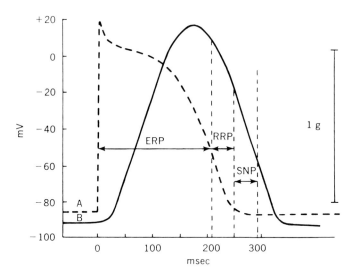

Fig. 14. Isolated papillary muscle of cat. Diagram showing the relationship between transmembrane action potential (A) and the isometric tension curve (B) as recorded from a small segment of muscle. Note that the relative refractory period (RRP) occurs during the early part of relaxation and is related to the membrane phenomenon and not to the contractile mechanism. [Reproduced, with permission, from Milnor, W.R.: Properties of cardiac tissue; in Mountcastle, V.B.: Medical Physiology; 14th ed.; vol. 2 (C.V. Mosby, St Louis, MO. 1980).]

cell from the interstitial fluid (through the sarcolemma, specially the T-tubules) and also from the sarcoplasmic reticulum (SR) which run longitudinally (fig. 15). The calcium ions bind to *troponin C* (which is the receptor of cardiac contractile protein) located in the thin *actin filament*. This binding is believed to move the tropomyosin laterally and uncover the binding sites on actin so that ATP can be split by *myosin ATPase* and provide the energy for the *myosin head* to interact with the binding sites of actin (fig. 16). ATP provides the energy for the movement of the crossbridges in the myosin molecules.

Cardiac Muscle Mechanics

The mechanical properties of heart muscle are not unlike that of skeletal muscle. The simplest way to study is to isolate the muscle and record the changes in length and force after stimulation. As a background, the reader should review such things as length-tension and

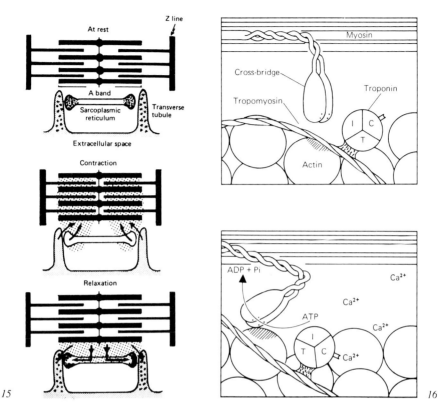

Fig. 15. Role of Ca^{2+} in muscle contraction. Ca^{2+} (black dots) is stored in the sarcoplasmic reticulum and is present in the transverse tubules which are part of the extracellular space. The spread of action potential to transverse tubules causes release of Ca^{2+} from tubules and sarcoplasmic reticulum into the cytosol of the muscle fiber. This triggers the processes shown in figure 16 which lead to the sliding of actin filaments (thin lines) on myosin filaments, and the Z lines move closer to each other. Ca^{2+} is then pumped back into the sarcoplasmic reticulum and transverse tubules by an active process and the muscle relaxes. [Modified from Ganong, W.F.: Review of Medical Physiology, 10th ed. (Lange Medical Publications, Los Altos 1981).]

Fig. 16. Action of Ca^{2+} initiating muscle contraction. Under resting conditions the troponin complex, which is attached to both actin and tropomyosin, holds the latter in a position that physically covers the binding site of actin, thereby preventing its interaction with the cross-bridge of myosin. Released Ca^{2+} combines with troponin C and produces a conformational change that detaches troponin from actin. This moves the tropomyosin and uncovers the binding site (shaded area). As a result, the cross-bridge of myosin can interact with the binding site of actin, causing contraction. Removal of Ca^{2+} reverses the process causing relaxation. [Reproduced with permission from Ganong, W.F.: Review of Medical Physiology, 10th ed. (Lange Medical Publications, Los Altos 1981).]

record the changes in length and force after stimulation. As a background, the reader should review such things as length-tension and force-velocity relationships of skeletal muscle in other texts. Explanations based on the sliding filament hypothesis must also be covered. We will briefly allude to studies on heart muscle.

The cardiac tissue most frequently used is the isolated papillary muscle. When a muscle is stimulated to contract and is prevented from shortening by an external opposing force, it develops force or tension. This type of contraction is described as *isometric* (same length). It has been found that the peak tension developed during contraction (active tension) varies with the length of the muscle before contraction begins (referred to as resting or initial length). At a certain initial length, the active tension developed is maximal. This length is designated as ℓ_0 or ℓ_{max} (optimal length). On either side of optimal length, with decreasing or increasing resting length, the developed or active tension declines (fig. 17). The rising phase of the curve to the left of ℓ_0 is described as the ascending limb and the falling phase to the right of ℓ_0 is the descending limb. The entire active tension curve is referred to as the *length-tension relationship* and is a fundamental property of muscle tissue. These observations have been partly (and incompletely) explained by the degree of overlap and interference of the crossbridges between actin and myosin filaments (for details refer to skeletal muscle studies).

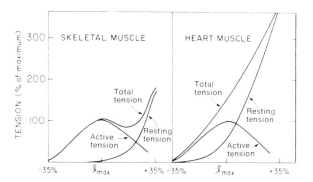

Fig. 17. Resting and active tension curves of heart muscle at different initial lengths as compared with that of skeletal muscle. Active tension curves are similar but resting tension of cardiac muscle is much greater than that of skeletal muscle even at lengths below ℓ_{max}. [Reproduced with permission from Katz, A.M.: Physiology of the Heart. (Raven Press, New York 1977).]

The length-tension relationship of heart muscle is of great importance in the intact heart for adjusting the volume of blood pumped out by each ventricle as related to the degree of filling of the ventricle before contraction (initial length). Normally, the ventricles operate on the ascending limb of the length-tension curve so that any increase in the filling of the ventricles increases the force of contraction and thereby the volume of the blood pumped out per beat. This relationship in the heart was first described by Otto Frank in 1895 and later elaborated by Ernest Starling in 1918, hence it is known as the *Frank-Starling relationship or mechanism.*

Another important aspect of muscle contraction is the relationship between velocity of shortening of the muscle and the force developed during contraction. For this purpose, muscle is stimulated and allowed to shorten against a constant load or an opposing force. This is designated as an *isotonic* contraction (same tension). Studies have shown that the rate of change of length at the beginning of shortening (dl/dt) varies with the magnitude of the load. The greater the load the slower the rate of change in length. The simplest model of muscle is to consider it as composed of at least two elements. The *contractile element* (CE) is taken to be in series with an elastic element called *series elastic* (SE). The experimental setup and the recording of changes in length and force are shown in figures 18 and 19. The muscle is attached rigidly at one end and a freely hanging weight fixed to the other. The weight constitutes the *preload* of the muscle (fig. 18a). The preload stretches the muscle to a certain degree, giving rise to the *resting* or initial length or tension of the muscle. Then a stop or support is provided that maintains the resting length of muscle when a second load is added, which constitutes the *afterload* (fig. 18a–c). Subsequently, the muscle is stimulated to contract. As shown in figures 18 and 19, the muscle at first does not shorten externally, but internally the contractile elements (CE) shorten and stretch the series elastic element (SE) an equal amount. This is an isometric contraction, since the external length of the muscle remains unchanged (figures 18 and 19d). Finally, the muscle begins to lift the two weights (both preload and afterload) and then relaxes, returning the weights back to the resting level. During the lifting phase, there is no further stretch of the series elastic element. The initial slope of the *external* shortening represents the initial velocity of shortening of the muscle (dl/dt).

The next step is to study the effect of increasing the afterload while keeping the *preload constant* (fig. 20). Note that as the afterload (or

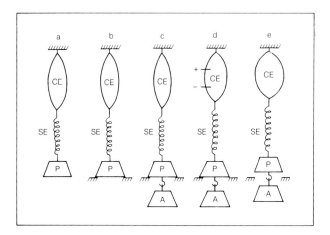

Fig. 18. Schematic representation of an afterloaded isotonic contraction of a muscle. In (a), a preload P is attached to the muscle which stretches the series elastic element (SE) to a certain degree. This creates a certain resting or initial length and tension in the muscle, depending on the magnitude of P. In (b), a stop or support is provided so that when an afterload A is added as in (c), resting length or tension will not change. In (d), the muscle is stimulated electrically and the shortening of the contractile element (CE) will first stretch the series elastic element (SE) without causing external shortening of the muscle. In (e), the external length of the muscle shortens, lifting both the preload P and afterload A (P + A is called the total load of the muscle). [Modified from Pollack, G.H.: Maximum velocity as an index of contractility in cardiac muscle: a critical evaluation. Circ. Res. *26:* 111 (1970). Reproduced with the permission of the American Heart Association, Inc.]

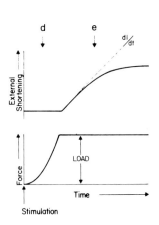

Fig. 19. Recording of changes in external length and force of the muscle shown in figure 18 after stimulation. Note that the first mechanical event after stimulation is stretch of the SE and a rise in tension or force without external muscle shortening (d). However, CE must shorten to stretch SE; this is the isometric phase and is followed by external shortening (e) of the muscle during which the force remains constant and equal to the total load of the muscle (P + A). This is the isotonic phase. The slope at the beginning of external shortening (dl/dt) represents the initial velocity of muscle shortening. [Reproduced, with permission, from Berne, R.M.; Levy, M.N.: Cardiovascular Physiology; 4th ed. (C.V. Mosby, St. Louis, MO. 1981).]

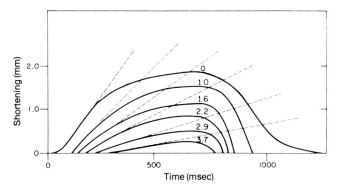

Fig. 20. Series of afterloaded contractions of a papillary muscle at a constant initial length. The numbers refer to the magnitude of the afterload in grams. Note the decreasing slope or velocity of shortening as the load is increased. Also, the onset of shortening is delayed (more time taken to stretch the SE elements) but the time from stimulation to maximal shortening remains unchanged. [Reproduced, with permission, from Berne, R.M.; Levy, M.N.: Cardiovascular Physiology; 4th ed. (C.V. Mosby, St. Louis, MO. 1981).]

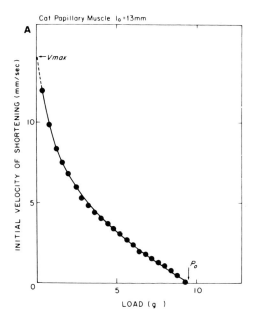

Fig. 21. The force-velocity relationship in the cat papillary muscle. P_o is the force developed when the muscle is unable to lift the total load (i.e., isometric contraction). [Reproduced, with permission, from Sonnenblick, E.H.: Implications of muscle mechanics of the heart. Fed. Proc. 21:975 (1962).]

total load) is increased, the initial velocity of shortening decreases. When the total load is too great, the muscle is unable to shorten *externally* to lift the load (not shown in the figure) and the contraction is entirely isometric (same length but tension changes). The maximum tension developed when external shortening is zero is referred to as P_0. If the initial velocity of shortening is plotted against the afterload or total load, with preload being constant, a roughly hyperbolic curve is obtained (fig. 21). This is the force-velocity curve of isolated heart muscle, which is very much like that of skeletal muscle. Extrapolation of the curve to zero load gives the theoretical maximal velocity of shortening (V_{max}), which is believed to indicate the maximal rate of cycling of the cross-bridges. A number of investigators have taken V_{max} as an index of the "contractility" or "contractile state" of the muscle. The contractile property of heart muscle has been described as the inotropic property (Greek, *inos* = fiber).

References

Cranefield, P.F.; Hoffman, B.F.: Electrophysiology of single cardiac cells. Physiol. Rev. *38:*41–67 (1958).

Fabiato, A.; Fabiato, F.: Calcium release from the sarcoplasmic reticulum. Circ. Res. *40:*119–129 (1977).

Frank, O.: On the dynamics of cardiac muscle [translated by Chapman, C.B.; Wasserman, E.: from Zeit. Biol. *32:*370–447 (1895)]. Am. Heart J. *58:*282–317; 467–478 (1959).

Hoffman, B.F.; Cranefield, P.F.: Electrophysiology of the Heart. (McGraw-Hill Book Co., New York 1960).

Katz, A.M.: Physiology of the Heart; pp. 137–159 (Raven Press, New York 1977).

Noble, D.: The Initiation of the Heartbeat; 2nd ed. (Oxford University Press, Oxford 1979).

Pollack, G.H.: Maximum velocity as an index of contractility in cardiac muscle. Circ. Res. *26:*111–127 (1970).

Ruch, T.C.; Patton, H.D.: Physiology and Biophysics; 19th ed.; pp. 1–72 (W.B. Saunders, Philadelphia 1965).

Sonnenblick, E.H.: Implications of muscle mechanics in the heart. Fed. Proc. *21:*975–990 (1962).

Starling, E.H.: The Linacre Lecture on the Law of the Heart. (Longmans, Green & Co., London 1918).

Weidmann, S.: Resting and action potentials of cardiac muscle. Ann. N.Y. Acad. Sci. *65:*663–678 (1957).

Weidmann, S.: Membrane excitation in cardiac muscle. Circulation *24:*499–505 (1961).

3 Spread of Cardiac Excitation

Unlike nerve and skeletal muscle tissues, cardiac muscle consists of heterogeneous muscular elements in which conduction velocity (table 2) and the size and duration of action potentials are not of the same magnitude but vary in different parts of the heart as already pointed out in Chapter 2.

The action potential from the true S-A pacemaker cells is conducted in the S-A node and then spreads into the surrounding atrial muscle in all directions. The velocity of conduction in typical atrial muscle is about 1 m/sec. To reach the A-V node, the classic concept was that there are no special pathways (fig. 22), but currently it is accepted by many that there are somewhat faster conducting pathways consisting of specialized muscle fibers. These are known as the *internodal* pathways and consist of three tracts called *anterior, middle* and *posterior* internodal which converge on the A-V node (fig. 23). The velocity of conduction in these tracts is about 1.5 m/sec.

On reaching the A-V node (which has three zones: atrionodal or AN zone, nodal or N zone, and nodal-His or NH zone), the excitation is slowed down to about 0.05 m/sec. This is explained on the basis of the *small* diameter of the fibers, which offers a high axoplasmic resis-

Table 2. Conduction velocity in different parts of the heart

	Velocity of conduction (m/sec)
Atrial muscle	1.0
Internodal fibers	1.5
A-V node	0.05
Bundle of His	1.0–1.5
Purkinje fibers	2.0–4.0
Ventricular muscle	0.3–0.4

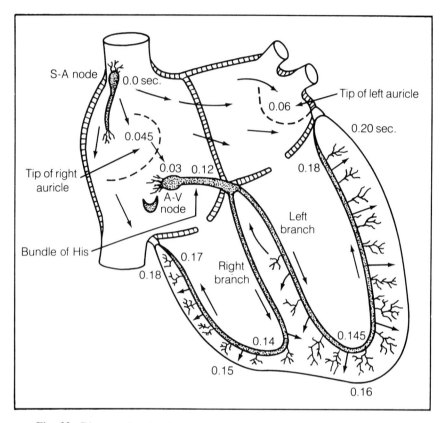

Fig. 22. Diagram showing the origin and spread of excitation over the mammalian heart. (Numbers indicate the approximate time of arrival at different points in the dog heart.) [Redrawn from Wiggers, C.J.: Physiology in Health and Disease; 5th ed. (Lea and Febiger, Philadelphia 1949).]

tance to ''local circuit'' currents. In the N-zone, the upstroke of the action potential has a reduced slope and amplitude. This is sometimes referred to as ''decremental'' conduction (in contrast to ''all or none'' conduction). The slow conduction in the A-V node is useful in that it allows the atrial contraction to be completed before the beginning of ventricular contraction. This serves to fill more blood into the ventricular cavity prior to systole and thus augment the stroke volume (blood pumped out per beat). At the same time it facilitates a more effective closure of the A-V valves. At the lower end of A-V node (NH zone) the velocity of conduction increases to that of the His bundle.

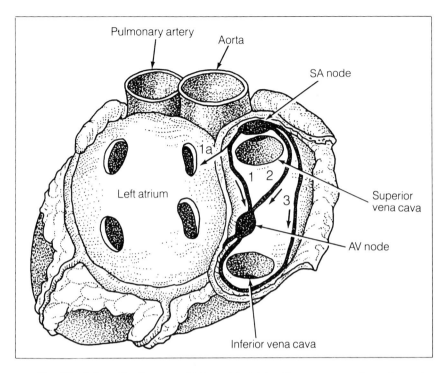

Fig. 23. Schematic diagram of internodal atrial pathways as seen from the posterior aspect of the heart, with the posterior wall of the right atrium removed. (1) Anterior internodal tract; (1a) branch of anterior internodal tract; (2) middle internodal tract; (3) posterior internodal tract. [Redrawn from Goldman, M.J.: Principles of Clinical Electrocardiography; 11th ed. (Lange Medical Publications, Los Altos 1982).]

From the A-V node the impulse reaches the *bundle of His* and its *two branches* (right and left) and then travels over the diffuse network of Purkinje fibers located under the ventricular endocardium (fig. 22) The velocity in the bundle of His and its branches is about 1.0–1.5 m/sec and in the Purkinje fibers it is about 2–4 m/sec. The rapid conduction in the Purkinje system permits *almost simultaneous* excitation and contraction of the various regions of each ventricular wall. This synchronous contraction plays an important role in ventricular dynamics, energetics and function as an effective pump.

The middle and apical parts of the interventricular septum are the earliest parts of the ventricles to undergo excitation (basal septum is depolarized later). The impulse spreads *from the left side* of the septum to the right side. The reason is that the right branch of bundle of His is slender and does not give off Purkinje fibers to the septum except at the apex, whereas the left branch gives off an extensive network of Purkinje fibers to the middle and apical septum. At the apex the impulses from the two branches travel against each other and cancel out by entering into the refractory period of the opposite side. Therefore, normally each ventricle is excited by its own bundle branch.

From the subendocardial Purkinje network the impulse proceeds through the ordinary muscle of both ventricles from endo- to epicardium (fig. 22). The muscle that is excited last is the epicardium of the left ventricle at its base. Conduction velocity over ordinary ventricular muscle is relatively slow, 0.3–0.4 m/sec. This partly explains why when one bundle branch is cut experimentally (or is destroyed by disease), the ventricle on that side is excited distinctly later than normal. Also, if the surface of a ventricle is stimulated during a nonrefractory period, causing a premature ventricular systole, such a contraction is *less synchronous* than normal.

Thus far we have outlined the sequence or spread of depolarization in the heart. The sequence of *repolarization* of cardiac chambers follows

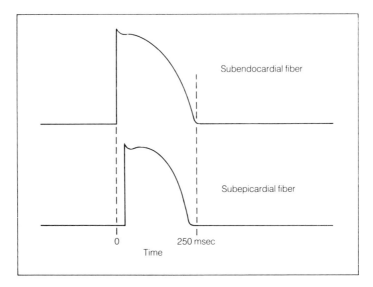

Fig. 24. Expected in vivo action potentials on the same time base from epi- and endocardial fibers of left ventricle based on in vitro studies.

that of depolarization in all parts except in the left ventricular myocardium. There is evidence that repolarization in the left ventricle proceeds from epi- to endocardium because the action potential persists longer in subendocardial compared with subepicardial fibers (demonstrated in isolated fibers, fig. 24).

If the bundle of His is cut across completely, or is functionally destroyed by lack of blood due to coronary disease, the ventricles stop receiving impulses from the S-A node and cease beating for a short period of time (complete A-V block). In man, this will cause fainting or convulsions as a result of cerebral ischemia (lack of blood flow; Stokes-Adams syndrome). Usually, however, the dormant intrinsic pacemaker of the bundle of His below the lesion, or in one of its branches, awakens and starts firing at a slow rate (30–40/min in man), being independent of the S-A node. This is called idioventricular rhythm. A person can live a sedentary life with this slow heart rate. During muscular activity, however, such individuals have a limited degree of cardioacceleration (increased cardiac sympathetic nerve activity and blood borne catecholamines from adrenals acting directly on the ventricular pacemaker) and hence, a limited increase in cardiac output and exercise tolerance.

Patients who do not develop adequate ventricular rhythm may be treated with "artificial pacemakers" at rates close to normal.

References

James, T.N.: The connecting pathways between the sinus node and the A-V node and between the right and left atrium of the human heart. Am. Heart J. *66*:498–508 (1963).

Scher, A.M.: Excitation of the heart; in Hamilton and Dow: Handbook of Physiology; sect. 2; Circulation; vol. 1; pp. 287–322 (American Physiological Society, Washington, D.C. 1962).

4 Principle and Recording of the Electrocardiogram

The electrocardiogram is a record of the changes in electrical potential on the surface of the body that occurs with the spread of each cardiac action potential. The spread of the action potential over various parts of the heart creates potential differences between the external surface of the "excited" (depolarized) and the "resting" (polarized) regions of the myocardium (externally, depolarized regions are electronegative with respect to resting polarized regions). Since the heart is surrounded by a conducting fluid and tissues with many electrolytes, *ionic currents* flow between the excited and resting regions through the surrounding medium. The body fluids and tissues are said to constitute a *volume conductor*.

To illustrate the principles involved in electrocardiography we shall resort to a study of very simple models. When a number of oppositely charged ions are held apart in a volume conductor at a very short distance they constitute what is called a *dipole layer* or simply a *dipole*. A dipole located in a volume conductor creates a flow of current in the form of ionic migration in the conducting medium, unlike the flow of loosely bound outer electrons called conducting electrons, in metals. By convention, current is said to flow from + to − (positive to negative). Actually, in liquids positively charged ions travel to the cathode (cations) and negatively charged ions to the anode (anions). This ionic migration, or separation, creates potential differences in the volume conductor which can be mapped out with an exploring (active) electrode connected to an indifferent electrode located at a great distance from the dipole (fig. 25). Thus the electrical "field" of the dipole may be mapped out. Lines of isopotential are found which are at right angles to the lines of current flow. The polarity of the exploring electrode is the same as that of the "nearer" surface of the dipole.

In a liquid with given electrical conductivity, *if the number of charges per unit area of a dipole (charge density) is constant,* the mag-

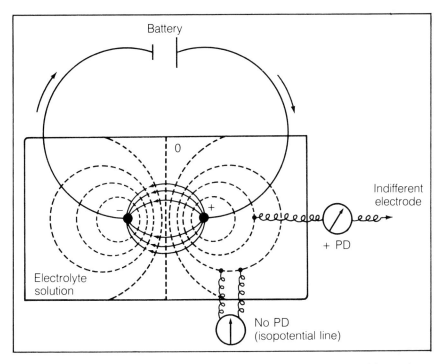

Fig. 25. Current flow and potential differences from a dipole located in a volume conductor.

nitude of the potential (voltage) at any point varies with three factors: (a) total area or size of the dipole layer, (b) orientation of the recording electrode with respect to the dipole, and (c) distance of the electrode from the dipole.

These three variables determine what is called the solid angle (Ω) between the recording electrode and the dipole layer.

$$\Omega = \frac{A}{r^2} \qquad A = \text{area of dipole "\textit{seen}" by the electrode} \left\{ \begin{array}{l} \text{factor (a) above} \\ \text{factor (b) above} \end{array} \right.$$

where r = distance between electrode and dipole seen (radius of curvature of arc) (see fig. 26).

Ω is maximal when dipole surface is perpendicular to the line from center of dipole to the electrode (fig. 27a). The solid angle decreases as the electrode moves away (less area is "seen" by the electrode, fig.

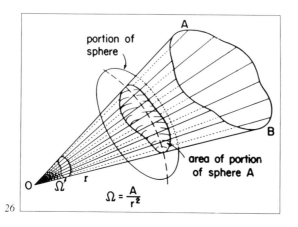

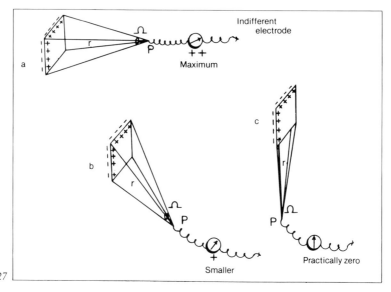

Fig. 26. Measurement of solid angle, omega, subtended by surfaces AB from O. Radii are drawn from O to all points on periphery of AB. This irregular cone is a solid angle. Omega is measured by inscribing a sphere of radius r with center at 0 and measuring the area, A, on the surface of sphere cut out by the irregular cone. Omega is defined as A/r^2. [Reproduced, with permission, from Woodbury, J.W.: Potentials in a volume conductor; in Ruch and Patton: Physiology and Biophysics; 19th ed. (W.B. Saunders, Philadelphia 1965).]

Fig. 27. Effect of area of dipole "seen" by recording electrode on solid angle. [Redrawn from Rushmer, R.F.: Cardiovascular Dynamics; 2nd ed. (W.B. Saunders, Philadelphia 1961).]

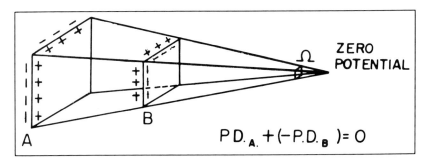

Fig. 28. Two dipole layers of equal charge density but of opposite polarity subtending the same solid angle at a given point neutralize each other. [Redrawn from Rushmer, R.F.: Cardiovascular Dynamics; 2nd ed. (W.B. Saunders, Philadelphia 1961).]

27b). No potential is recorded when the electrode is equidistant from the + and − charges (fig. 27c). The exact contour or shape of the dipole layer is immaterial. It is the size of the solid angle that matters. The polarity of the potential depends on that of the pole facing the electrode. If the near surface is positive, the recorded potential is positive and vice versa. Usually positive is recorded upward.

What happens when two dipole layers of *opposite* polarity subtend the same solid angle at a given point? The angles add algebraically and no potential will be recorded (fig. 28). This is the case with a *resting* cell or resting heart muscle mass that is polarized equally over its entire surface and is located in a volume conductor (fig. 29).

When such a cell is stimulated at a point *away* from the recording electrode, a positive potential is recorded during the spread of depolarization (fig. 29). It is followed by an opposite deflection during the period of repolarization which follows the path of depolarization. The shape of the second deflection depends on the time sequence of repolarization. If the time sequence is exactly the same as depolarization, the second deflection will be a mirror image of the initial (in the model this is assumed to be the case). This, of course, is not true in heart muscle (see fig. 7).

Using an elongated cell located in a volume conductor, it can be shown that the potential recorded at an external point depends *only* on the solid angle of the *boundary* between the excited and resting regions (fig. 30). This boundary constitutes the dipole. This principle applies regardless of the shape of the cell or muscle mass.

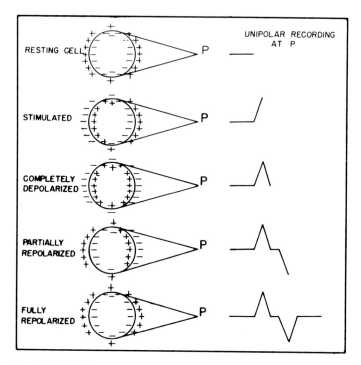

Fig. 29. Unipolar recording of action potential from a cell or cell mass located in a volume conductor (the magnitude and rate of depolarization and repolarization are assumed to be the same at different points on the membrane). Note the biphasic recording under such circumstances.

In the case of the human heart, the dipole (consisting of the boundary between active and resting regions) is not stationary but is travelling over a very complex geometric muscle mass and is continuously changing its total size, shape, and speed of travel. Also, the rate of repolarization is different from that of depolarization. As a result, a very complex recording is obtained (the electrocardiogram).

The early studies in man were done by Einthoven not by a unipolar method as shown in the models, but by a more complex *bipolar* method (adding more variables). Electrodes were applied on two points on the surface of the body. Both of these points undergo changes in potential with each heartbeat and the record obtained consists of the *differences* in potential between electrodes that are both "active." Einthoven introduced the three standard limb leads (bipolar):

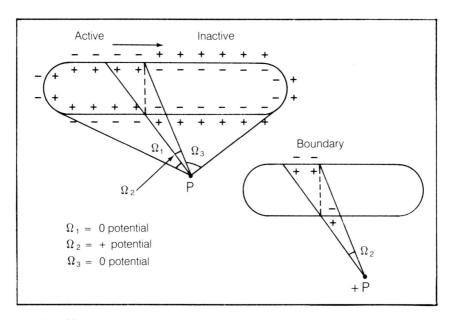

Fig. 30. Potential at an external point P due to rising phase of action potential in an elongated excitable cell located in a volume conductor. The junction between the active and inactive regions is assumed to have an abrupt reversal of polarity. Total solid angle at P is divided into 3 portions: Ω_1, Ω_2, and Ω_3. The potential at P due to solid angles Ω_1 and Ω_3 is zero, since the nearer and farther membrane dipoles have equal solid angles but opposite polarity. However, in Ω_2 both the nearer and farther membrane dipoles have the same polarity (positive) which are additive. The figure on the right shows that potentials at an external point depend only on the solid angle of the *boundary* between the excited and the resting regions of the cell or tissue. [Redrawn from Wood-bury, J.W.: Potentials in a volume conductor; in Ruch and Patton: Physiology and Biophysics; 19th ed. (W.B. Saunders, Philadelphia 1965).]

Lead I   = Difference of potential between left arm and right arm (LA-RA) (fig. 31).
Lead II = Difference of potential between left leg and right arm (LL-RA).
Lead III = Difference of potential between left leg and left arm (LL-LA).

By convention, upward deflection in Lead I is recorded when the right arm is relatively *electronegative*. Upward deflection in Lead II and III is recorded when the *respective arms* are *negative* (fig. 31).

In 1934, Wilson introduced the so-called *unipolar* leads. The wires from the three limbs were connected together after passing through 5000-ohm resistances. This junction, called the *central terminal*, was shown to have approximately zero potential during the spread of cardiac action

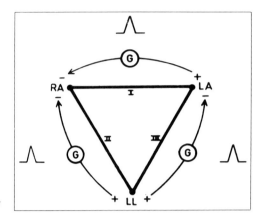

31

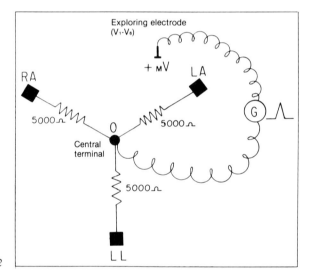

32

Fig. 31. Polarity of electrodes in the three standard limb leads to produce galvanometer deflections above the isopotential line (agreed by convention).

Fig. 32. Wilson's central terminal lead to obtain approximately zero potential during cardiac cycle (unipolar or V lead). Deflection is upward when exploring electrode is positive.

potential. Hence, it constituted the *indifferent* electrode (fig. 32). Such leads are called V leads (voltage) and the "active" electrode may be located on one of the three extremities (VR, VL, VF) or may be on the chest wall near the heart (V_1, V_2, V_3, V_4, V_5, V_6) (see fig. 33).

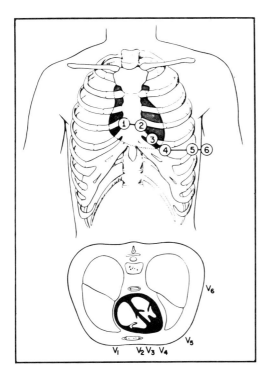

Fig. 33. The positions on the chest to record the unipolar precordial leads, V_1 – V_6. V_1 is to the right of the sternum at the 4th intercostal space. V_2 is to the left of the sternum at the 4th intercostal space. V_4 is in the 5th intercostal space on the midclavicular line. V_3 is between V_2 and V_4. V_5 is in the 5th intercostal space on the anterior axillary line. V_6 is in the 5th intercostal space on the midaxillary line. The lower figure shows the positions of the leads in the transverse plane of the chest. [Reproduced, with permission, from Scher, A.M.: Electrocardiogram; in Ruch and Patton: Physiology and Biophysics; 20th ed.; vol. II; Circulation, Respiration and Fluid Balance. (W.B. Saunders, Philadelphia 1974).]

In 1942, Goldberger modified the Wilson technique so as to increase the amplitude of the unipolar limb leads by about 50% (amplitude of VR, VL, and VF were found to be too low for clinical use). This was done by breaking the connection to the central terminal from the limb on which the exploring electrode is placed. These recordings are called *augmented* limb leads; AVR, AVL, AVF (AVR = 3/2 VR). Obviously, these are not truly unipolar leads. Later it was found that the 5000-ohm resistances could be eliminated with similar results in the augmented leads.

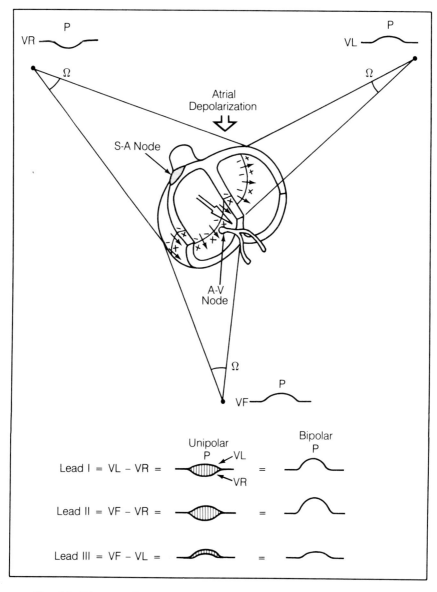

Fig. 34. Diagram of atrial depolarization to illustrate the principle of a dipole (boundary) in a volume conductor. The solid angle and the polarity of electrode deflections, using the unipolar limb recordings (VR, VL, VF or AVR, AVL, AVF). Atrial dipole has a negative potential (excited region) nearer to the right arm and a positive potential (resting) nearer to the left arm and left leg. Hence, the P wave is downward in VR (or AVR) and upward in VL and VF (or AVL and AVF).

Currently, the electrocardiograph has a switch to make the proper connections for standard bipolar Leads I, II and III; AVR, AVL and AVF; and unipolar chest leads V_1 through V_6 (12 in all).

Figure 34 illustrates the application of the concept of solid angle and electrode polarity to the atrial depolarization wave (P-wave) in unipolar limb leads (VR, VL, VF). Note that the standard bipolar limb leads may be derived from the unipolar limb leads (VR, VL, VF) as follows:

Lead I = VL − VR; Lead II = VF − VR; Lead III = VF − VL.

The above method of recording is described as *scalar* electrocardiography in which potential differences have magnitude but the direction (polarity) is in one axis only. This is in contrast to *vector* ECG in which cardiac potentials are described as having magnitude and direction in a *plane* surface (two-dimensional) or in a *cube* (three-dimensional or spatial).

Physiologic Basis of the ECG

Each heart beat is accompanied by a series of deflections, labeled by Einthoven as P, Q, R, S and T. Sometimes a U wave is present (fig. 35). Not all of these deflections are present in all leads and there are considerable variations in their amplitude in different leads (fig. 36). The electrical potentials are more or less of the order of a millivolt. This is much less than the transmembrane action potentials of a single cardiac muscle cell, which is about 120 mV. *Reason:* other things being equal, voltage at any point in a volume conductor varies inversely with the square of the distance of the point from the dipole ($\Omega = A/r^2$). By the same token, the deflections in the chest leads have a higher voltage than those in the limb leads (fig. 36). [*Note:* The electroencephalographic (EEG) potentials of the brain as recorded from the scalp are of the order of μV which require much more amplification for recording]. By convention the time scale on the abscissa is 1 mm = 0.04 sec (paper speed 25 mm/sec) and voltage scale on the ordinate is 1 mm = 0.1 mV; a heavier line is present at every 5 mm on both axes (0.2 sec and 0.5 mV, respectively).

The P-wave. It is produced by the spread of *depolarization* over the musculature of the *two atria*. The end of P represents the depolarization

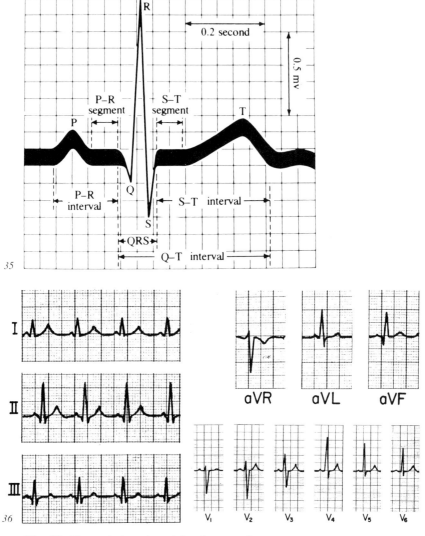

Fig. 35. The typical tracing of the electrocardiogram. The deflections are labelled and the segments and intervals are indicated. The standard time scale and voltage scale are also shown. [Reproduced, with permission, from Abel, F.L.: Heart and circulation; in Selkurt: Basic Physiology for the Health Sciences; 2nd ed. (Little, Brown and Company, Boston 1982).]

Fig. 36. The normal scalar ECG taken in 12 leads. There are many normal variations, depending on the position of the heart in the chest, including changes with the respiratory cycle. [Reproduced, with permission, from Guyton, A.C.: Textbook of Medical Physiology; 5th ed. (W.B. Saunders Co., Philadelphia 1976).]

of atrial muscle which is farthest from the SA node. This is somewhere on the left atrial appendage and not at the AV node as some might think. The average duration of the P-wave is about 0.08 sec and its amplitude is about 0.15 mV.

The isopotential after P, called the *P-R segment*, registers while the depolarization is passing over the lower part of the AV node, the bundle of His, and the upper part of bundle branches. The reason why no voltage is recorded on the ECG is that the tissues are too small to cause a detectable potential on body surface. In other words, the dipole size is too small and therefore the solid angle is too small. Also, the size of the action potential and its slope are low. During this period, the musculature of the two atria remains in the plateau phase. The duration of the P-R segment is about 0.08 sec.

The remaining Q, R, S and T deflections are produced by the spread of the action potential over the muscle mass of the *two ventricles*. Whether or not the electrical activity of all ventricular muscle cells registers in the recording is uncertain. There is evidence to believe that electrical activity in some cells does not appear on the Q, R, S and T deflections, eg, infarction of the left ventricular subendocardium associated with death of these cells may not be detected by ECG (electrocardiographically silent area).

The initial ventricular deflection, called the *QRS complex*, is due to the *rapid* spread of depolarization over the two ventricular muscle masses. The beginning of Q or R represents the depolarization of the middle part of the ventricular *septum on the left side*, which is excited first. The end of S represents depolarization of the epicardial surface of the left ventricle at its base (last region to be depolarized). The duration of QRS (from the beginning of Q to the end of S) is about 0.08 sec and its amplitude varies in different leads from 0.5 to 1.5 mV.

The *isopotential* after the QRS complex is called the *S-T segment* (fig. 35). It represents the period during which the two ventricles are in the "plateau" phase where the muscle cells are almost equally depolarized for an appreciable period of time. Its duration is about 0.12 sec.

The *T-wave* represents the period of repolarization of the two ventricles. It is of smaller amplitude and longer duration than the QRS complex because the *rate* of repolarization of ventricular muscle cells is slower than the rate of depolarization. The amplitude of T is usually greater than that of P (larger muscle mass) and its duration is longer than that of P (about 0.16 sec). Because the T-wave is upright and has the same polarity as the QRS, it was suspected by early electrocardiologists that repolarization of the ventricles proceeds from the epicardium to the endocardium. This is explained by the *longer duration* of action potential in the subendocardial fibers as shown by in vitro studies (inborn, see fig. 24, Ch. 3). It may also be *partly* caused by the greater pressure in subendocardial fibers during systole, particularly in the left ventricle.

During the T-wave, the ventricles are relatively refractory to stimuli which arise from foci outside of S-A node (ectopic foci). If the ventricles are stimulated at a certain critical moment during this period (repolarization), described as the "vulnerable" period, they undergo fibrillation. Fibrillation of ventricles is an incoordinate contraction of ventricular syncytium resulting in failure to develop adequate pressure and pumping of blood. This may occur after myocardial infarction, from electrical currents passing through the body (electrocution) or from toxic doses of certain drugs (digitalis).

If the T-wave is due to repolarization of the ventricles, where is the repolarization wave of the atria (T_a-wave) on the ECG? This deflection is small due to the small mass of atrial tissue and the slower rate of repolarization and is lost in the much larger potential of the QRS complex. In complete A-V block, one may see a small T_a-wave whose polarity is opposite to that of the P wave (in the atria, the sequence of repolarization is the same as that of depolarization).

P-R Interval. This is measured from the *beginning* of P to the beginning of Q or R (fig. 35). Normally in adults it varies from 0.12 to 0.20 sec. If it is longer than 0.2 sec, it indicates an abnormal delay in the conduction from the S-A node to the middle of the ventricular septum. Usually the delay is at the region of the A-V node.

Q-T Interval. This interval is measured from the *beginning* of Q or R to the *end* of T (fig. 35). This period is referred to as the "electrical systole" of the ventricles. Its average duration is about 0.36 sec.

Abnormalities in the Contour of ECG Deflections

For any given lead, abnormalities in the shape or contour of the deflections may be due to one or more of the following causes:
1. Change in the *site* of the pacemaker (located outside of S-A node—referred to as ectopic). This will necessarily change the path of spread of the action potential.

2. Change in the *speed* of conduction of the impulse in localized areas. Hence, summation with respect to time will be different from normal and result in a change in the shape of deflection, eg, areas of mild ischemia (lack of blood supply) of ventricular myocardium will produce this.

3. Death of ventricular muscle (cardiac *necrosis* or *infarction* from lack of bloody supply). These muscle cells lose their resting and action potentials. If a sufficiently large number of cells is involved, the net result of the electrical activity in the remaining cells will be different from normal. Usually the change is in the *S-T segment* or in the T-wave. S-T segment tends to lie distinctly above or below the isopotential line.

His Bundle Electrogram (HBE)

It is possible to study the conduction along the A-V node and the bundle of His and its branches in more detail by means of a catheter that

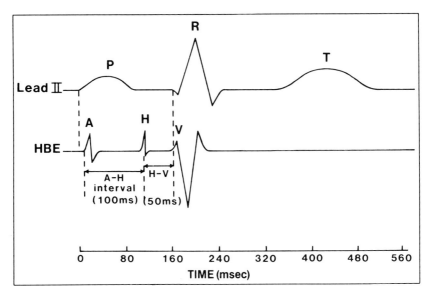

Fig. 37. Diagram illustrating the normal relationship of the His bundle electrogram (HBE) to the deflections of the scalar electrocardiogram (Lead II). Note that a major part of the P-R interval in the ECG is due to the slow conduction through the A-V node (represented by the A-H interval in the HBE).

has multiple recording leads at its tip. The catheter is passed through a vein into the right atrium and ventricle so that the tip lies close to the A-V node and the ventricular septum. Such direct recordings are known as electrograms. In this case it is referred to as the His bundle electrogram (HBE). It is recorded simultaneously with surface electrocardiographic leads at high paper speed—100 mm/sec or more as compared with the normal ECG speed of 25 mm/sec.

There are three significant complexes in the HBE. The first is due to the activation of the lower atrial muscle and is called the A deflection. The second—called the H deflection—is due to the electrical activity of the bundle of His and is very short lasting. The third is a complex due to the activity of the ventricular septum, the V deflection (fig. 37). The interval between the beginning of A and the peak of H represents essentially the conduction through the A-V node (A-H interval). The interval between H and the beginning of V (or R on the ECG) measures the conduction time from the bundle of His to the activation of the ventricular septum (usually at about its middle). In various disturbances of conduction in these tissues, HBE is useful in facilitating a more accurate diagnosis and serves as a guide to their therapy.

References

Burch, G.E.; Winsor, T.: A Primer of Electrocardiography; 6th ed. (Lea and Febiger, Philadelphia 1972).

Goldberger, E.: A simple, indifferent, electrocardiograph electrode of zero potential and a technique of obtaining augmented, unipolar, extremity leads. Am. Heart J. 23:483–492 (1942).

Goldman, M.J.: Principles of Clinical Electrocardiography; 11th ed. (Lange Medical Publications, Los Altos 1982).

Rushmer, R.F.: Cardiovascular Dynamics; 2nd ed.; pp. 240–245 (W.B. Saunders, Philadelphia 1961).

Wilson, F.N.; Johnston, F.D.; Macleod, A.G.; Barker, P.S.: Electrocardiograms that represent the potential variations of a single electrode. Am. Heart J. 9:447–458 (1933–34).

5 Cardiac Vector Analysis

The potential differences that are generated in the heart during the spread of cardiac action potential may be represented by a series of spatial vectors which have a continuously changing direction and magnitude. At any one instant, there are numerous small vectors in different directions, but one can represent the *average* direction and the *net* potential developed by what is called the *mean instantaneous vector*. By convention, an arrow is used to represent the instantaneous vector whose direction may be taken to represent the direction of current flow *through the myocardium*. Thus, the cardiac vector is directed from excited regions of heart muscle (surface relatively negative) to inexcited or "resting" regions (surface positive) (fig. 38). This representation appears contradictory to the convention of current flow which is from positive to negative. The length of the arrow represents the overall voltage generated.

In the body the spread is very complex on account of the fact that the heart is a heterogeneous organ, has a very complicated geometry, and is located in a nonuniform volume conductor. Cardiac vector analysis began by Einthoven who introduced the concept of the "equilateral triangle." This was based on the assumptions that (a) the extremities form an equilateral triangle in the coronal (frontal) plane with the apex of the triangle being directed toward symphysis pubis, and (b) the heart lies at the center of this triangle and is surrounded by a volume conductor having a uniform conductivity. Although these assumptions cannot be entirely correct, yet vector analysis by the Einthoven method is accurate enough to be useful clinically.

The Graphic Method of Analysis. The standard (bipolar) limb leads are taken to represent the three sides of the equilateral triangle. The midpoint of each side of the triangle represents zero potential and perpendiculars to these points meet at the center of the triangle. Conven-

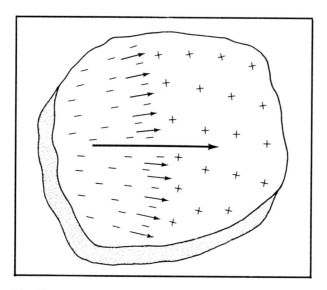

Fig. 38. Partly depolarized sheet of cardiac muscle. Note the small instantaneous vectors through the myocardium in different directions and the mean or summated instantaneous vector (large arrow) directed from the excited to the resting region of the muscle.

tional polarity is marked on each side of the triangle to indicate the upward (+) or downward (−) deflection in each of the three leads (fig. 31).

Simultaneously recorded standard limb leads are secured. Suppose it is desired to find the vector of the QRS complex at any one instant. Simultaneous instants on any two leads (eg, lead I and lead III) of QRS are noted on the tracing and the values of these points (in arbitrary units) are plotted on the corresponding sides of the triangle. From these points, perpendicular lines are drawn. Their point of intersection represents the head of the arrow, which is drawn from the center of the triangle. This arrow represents the "instantaneous" vector at the selected instant of QRS. The procedure is repeated for different instants on the QRS complex, obtaining a series of instantaneous vectors. Each instant will have a vector of different direction and magnitude, indicating the spread of ventricular depolarization. If the heads of all vectors are joined together, a loop—called a vectorcardiogram (or vector loop)—is obtained. The direction of movement of vector is indicated on the loop (fig. 39). This loop represents the vector in the *frontal* or *coronal* plane.

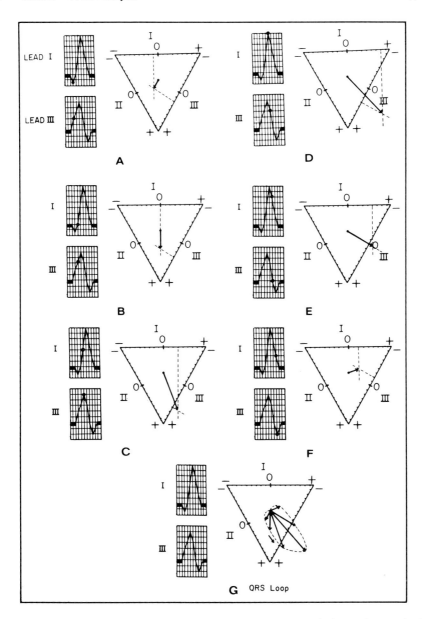

Fig. 39. Graphic derivation of QRS vector loop in the coronal plane using standard limb leads (recorded simultaneously) and the Einthoven triangle. The P loop and the T loop are derived in a similar fashion. [Arranged and reproduced with permission from Burch, G.E.; Abildskov, J.A.; Cronvich, J.A.: Spatial Vectorcardiography. (Lea and Febiger, Philadelphia 1953).]

Since, in the ECG, there are three deflections—P, QRS, and T—each preceded by an isopotential line, there are normally three loops with each heart beat. One loop is for the P-wave, one for the QRS, and one for the T-wave. Of these, the largest is that of QRS because it has the highest voltage. In the frontal plane this vector usually, but not always, travels in a counterclockwise direction. The graphic method is tedious and has given way to the more practical oscilloscope for recording vector loops.

Oscilloscopic Method. This is a much simpler method of securing vector loops. Using the oscilloscope, lead I is fed into the X-axis and lead II or III into the Y-axis. The result is a recording of the vector loops with each heart beat. This procedure will give the loops as "seen" on the frontal or *coronal plane* because the placement of electrodes on the extremities taps the cardiac voltages in that plane.

If the cardiac vector is to be studied in three dimensions, the electrodes may be placed on the chest along three axes (lateral, vertical and anteroposterior). An example is the so-called cube system of electrode placement (fig. 40). With this type of recording one can visualize the *spatial* orientation of the cardiac vector (spatial vectorcardiogram). The

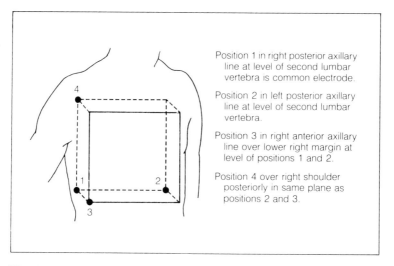

Position 1 in right posterior axillary line at level of second lumbar vertebra is common electrode.

Position 2 in left posterior axillary line at level of second lumbar vertebra.

Position 3 in right anterior axillary line over lower right margin at level of positions 1 and 2.

Position 4 over right shoulder posteriorly in same plane as positions 2 and 3.

Fig. 40. Diagram showing the cube system of electrode placement in spatial vectorcardiography. [Reproduced, with permission, from Burch, G.E.; Abildskov, J.A.; Cronvich, J.A.: Spatial Vectorcardiography. (Lea and Febiger, Philadelphia 1953).]

clinical usefulness of these studies is still uncertain and there is at present no standardized placement of electrodes on the chest.

Mean Electric Axis of QRS

A great simplification of cardiac vectors is to find the general direction of instantaneous vectors using the standard limb leads. This will indicate what is called the *mean electric axis* of the ventricles. It is commonly done for QRS because it is the largest of the three loops. Using the graphic method of analysis, the algebraic sum of R (+) and S (−) deflections in any two leads are plotted. These points are then located on the respective sides of the Einthoven triangle. Intersection of perpendiculars from these points marks the head of the arrow (fig. 41).

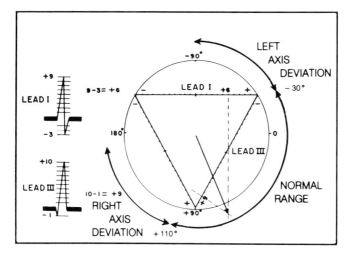

Fig. 41. The mean electrical axis is computed from two of the three standard limb leads (eg, leads I and III). The sum of the downward deflections is subtracted from the sum of the upward deflections. For example, the vertical height of the R wave above the baseline is measured in millimeters (+9 mm in lead I). The total amplitude of the downward deflections (−3 mm in lead I) is added algebraically to the height of the R wave (+9) and leaves a net value of +6. At a point 6 units toward the plus sign on the lead I line, a perpendicular is erected. The net amplitude of upward and downward deflections in Lead III is +9 (+10 − 1). A perpendicular erected 9 units toward the plus sign on lead III is extended to intersect the perpendicular from lead I. An arrow drawn from the center of the triangle to the intersection of these two perpendicular lines is the mean electrical axis. [Modified from Rushmer, R.F.: Cardiovascular Dynamics; 2nd ed. (W.B. Saunders, Philadelphia 1961).]

Normally, the mean electric axis is directed between $-30°$ and $+110°$. If the axis is directed between $-30°$ and $-90°$ (above horizontal), there is said to be left axis deviation. If the axis is between $+110°$ and $+180°$, there is right axis deviation (fig. 41).

Some of the deviations of the mean electric axis of QRS are caused by:

1. Altered position of the heart in the chest (eg, pregnancy, fluid in the peritoneal cavity or ascitis, and so forth).
2. Hypertrophy of ventricular muscle on one side, or dilation of a chamber.
3. Abnormal direction of spread of ventricular excitation (ectopic focus in ventricles or bundle branch block).
4. Infarction of ventricular muscle.

Reference

Burch, G.E.: Abildskov, J.A.; Cronvich, J.A.: Spatial Vectorcardiography. (Lea and Febiger, Philadelphia 1953).

6 The Cardiac Cycle

Contraction of the heart imparts energy to the blood contained in the chambers, causing energy gradients and flow. Most of the energy is in the form of pressure (potential energy) but a certain amount is kinetic ($\frac{1}{2}$ mv^2). Energy (chiefly pressure) differences on the two sides of the cardiac valves close or open the valves, depending on their anatomic arrangement, which permits unidirectional flow. They act like reciprocating valves. Contraction of heart muscle is called *systole;* relaxation and rest period are *diastole.* The sequence of mechanical events during one heart beat is called the *cardiac cycle.*

The description of the cardiac cycle can begin with any phase, but it is convenient to start with ventricular systole, which begins when each ventricle—in a resting person—contains about 150 ml of blood (end-diastolic volume, EDV) at a pressure of a few millimeters of mercury. This volume represents the *preload* of the ventricle, which determines the initial length of the muscle fibers. On contraction, ventricular pressure rises and immediately exceeds atrial pressure, closing the A-V valves (fig. 42). Now the ventricles are completely closed chambers and the pressure rises very rapidly. The contraction is said to be *isovolumic;* the heart becomes more spherical without changing volume. The slope of the pressure rise (dp/dt) indicates the velocity of contraction; its maximum has been used as an index of "contractility" of the ventricular myocardium. When the ventricular pressure exceeds the *diastolic* arterial pressure, the semilunar valves open and blood is ejected into the respective artery. The arterial pressure during ejection represents the *afterload* of the ventricle, except in stenosis of the semilunar valves. The flow increases first, rapidly reaches a peak, and then declines to zero (fig. 43). The volume ejected is about 70 or 80 ml/ventricle in a supine resting person and is the *stroke volume* (SV). This phase of contraction is the "ejection" phase and the muscle fibers shorten—isotonic! Actually, the contraction is "auxotonic," which refers to shortening

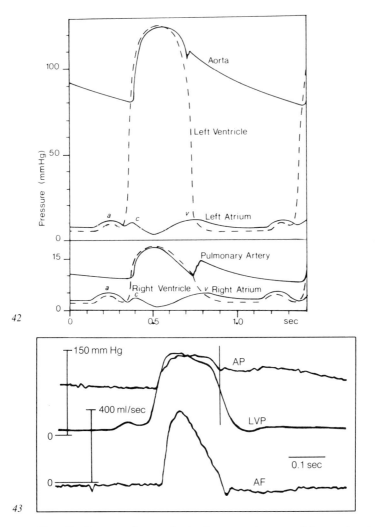

Fig. 42. Pressure changes in the large arteries, ventricles, and atria during one cardiac cycle lasting 1 sec (60 beats/min). [Modified from Burton, A.C.: Physiology and Biophysics of the Circulation; 2nd ed. (Year Book Medical Publishers, Chicago 1972).]

Fig. 43. Record of left ventricular pressure and aortic pressure and flow from a conscious dog under resting conditions. AP = pressure in proximal ascending aorta; LVP = left ventricular pressure; AF = aortic blood flow. Note that the left ventricular pressure toward the end of ejection (see aortic flow record) is lower than that in the aorta. [Reproduced, with the permission of the American Heart Association, from Noble, M.I.M.: The contribution of blood momentum to left ventricular ejection in the dog. Circ. Res. 23: 662 (1968).]

with increasing tension. The blood remaining in each ventricle at the end of systole is called end-systolic volume (ESV) or *residual volume*, and is about equal to the stroke volume (70–80 ml) in the supine position. The ratio of stroke volume to EDV is the *ejection fraction* (SV/EDV), which normally is about 50%–60%. It has also been used as an index of ventricular contractile function.

During the ejection of blood, the aortic pressure rises, reaches a peak (peak systolic), and then declines towards the end of ejection. The decline is due to the fact that the volume of blood leaving the aorta is greater than that entering from the left ventricle. In recent years, accurate studies of ventricular and aortic pressures during ejection have shown that the ventricular pressure during the second half of ejection is *lower* than that of the aorta despite the fact that flow is from the ventricle to the aorta (fig. 43). This apparent paradox is explained by the fact that flow is caused by the total *energy* difference between two points and total energy consists not only of pressure but also of kinetic energy (see Bernoulli's equation in Chapter 13). The total energy of blood in the ventricle is greater than its total energy in the aorta during this period. The pressure difference (better energy difference) between the left ventricle and the aorta during ejection is *very small,* because the aortic orifice is large (about 3 cm^2) and offers very little resistance to flow (see Poiseuille's law, Chapter 13). During ejection, the base of the heart moves toward the apex (fig. 44); all dimensions of the heart diminish. The semilunar valve leaflets do not make contact with the arterial walls

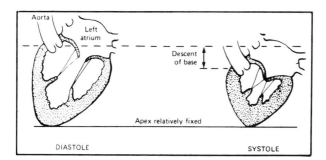

Fig. 44. Diagram of the left ventricle showing the descent of the base of the ventricle toward the apex during ejection. [Reproduced with permission from Sokolow, M.; McIlroy, M.B.: Clinical Cardiology; 3rd ed. (Lange Medical Publications, Los Altos 1981).]

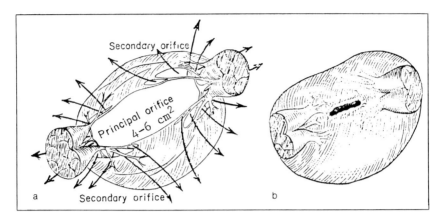

Fig. 45. (a) Diagram of the normal mitral valve as seen from below. The orifices have a large area which offers a low resistance to blood flow. (b) Diagram of the stenotic (narrowed) mitral valve as seen from below. The principal orifice has been narrowed extremely and the secondary orifices completely obliterated by adhesions of the valve leaflets and chordae tendineae. This opening offers a very high resistance to blood flow. [Reproduced with permission from Bonnabeau, R.C.; Stevenson, J.E.; Edwards, J.E.: Obliteration of the principal orifice of the stenotic mitral valve: a rare form of "restenosis." J. Thorac. Cardiovasc. Surg. *49:* 264 (1965).]

but take a midposition as a result of eddy formation behind them. In the aorta this mechanism safeguards the patency of the orifices of the coronary arteries.

When ventricular relaxation or *diastole* begins, aortic pressure drops sharply (called incisura) and the semilunar valves close immediately. The ventricles are again closed chambers and the pressures fall rapidly as relaxation proceeds. This is the *isovolumic relaxation* phase. During these periods, the A-V valves are closed and the atria are filling with blood from the veins, but atrial pressure rises *slowly* because of the relatively thin walls and the great distensibility of the atria. When ventricular pressure drops below the atrial pressure, the A-V valves open and blood flows into the ventricles due to the energy gradient. The flow is rapid at first, slowing down as the atria empty and the ventricles fill. The slow filling phase is sometimes referred to as the period of *diastasis*. There is evidence that the relaxing ventricle exerts a very slight sucking force that tends to lower the ventricular pressure and help filling. However, the major factor for ventricular filling is atrial pressure during atrial diastole. This pressure is ultimately derived from the energy imparted to the blood by the contraction of the opposite ventricle. Toward the end of ventricular diastole, systole of the atria takes place, raising the pressure in both chambers and driving some additional blood into the ventricles. The contribution of atrial systole is estimated to be about 10%–30% of the filling during one cycle (20% of 75 ml is about 15 ml).

The mean A-V pressure difference throughout the three filling phases is known as the ventricular filling pressure and is normally a few millimeters of mercury. At rest, this low pressure gradient drives about 70–80 ml of blood in a short period of time (about 0.45 sec) because the A-V orifices are large (4–6 cm^2) and offer little resistance to flow (fig. 45). There are two basic factors that determine ventricular filling per cycle: (a) mean A-V pressure differences during the three filling phases and (b) duration of the three phases. Within limits, the longer the duration, the greater the filling. Slow heart rate increases the duration of diastole and increases the stroke filling, up to a limit. Increased heart rate shortens both systole and diastole, but diastole is shortened more, particularly the period of diastasis which may be completely eliminated at rapid heart rates. During muscular exercise, the increase in heart rate does not cause a reduction in filling-per-cycle, even though diastole has shortened. The stroke volume or flow per cycle is usually maintained at the resting level or may even increase slightly. This must be due to a slight increase in atrioventricular pressure gradient, but studies have not demonstrated a statistically significant increase in right atrial pressure. The reason is probably related to the fact that the A-V orifice is large and a very slight increase in the pressure gradient can maintain filling-per-cycle at the resting level (70–80 ml) or even increase it, but present day methods of measurement do not detect such minor pressure increases (error of method is greater). On the other hand, if heart rate increases moderately (eg, 120 beats/min) in a *resting individual*, there is a proportional reduction of stroke filling (or stroke volume) so that the output per minute remains unchanged. However, if the rate is *very fast*, say 160/min or more in a *resting individual* (eg, paroxysmal atrial tachycardia), the period of diastole may be so short that there is very little filling per cycle. The stroke volume may be so low that the output per minute drops below the normal value (*ca* 5 L/min) despite the rapid rate. Blood pressure drops and blood accumulates on the venous side, raising atrial and venous pressures (in both the pulmonic and systemic circuits).

Coming back to the cycle, note that the end of atrial relaxation is soon followed by the *beginning* of ventricular systole.

The pressure developed by the left ventricle is much greater than that of the right ventricle because the resistance of systemic vessels is much greater than that of the pulmonic (fig. 42). To develop the higher pressure in the left ventricle, the muscle wall is much thicker (three times thicker in humans). If there is a ventricular septal defect, the blood will obviously flow from left to right. Although the mitral valve closes slightly before the tricuspid, the low diastolic pressure in the pulmonary artery permits the pulmonary valve to open before the aortic valve opens. These findings explain why the period of isovolumic contraction of the right ventricle is of shorter duration than that of the left.

The atrial pressure recording shows three major elevations. The first, called the *a-wave,* is caused by the contraction and relaxation of atria, which precede ventricular systole (fig. 42). The next is the *c-wave,* which is caused by the bulging of the A-V valves toward the atria during the isovolumic contraction of the ventricles. This is followed by a fall in pressure (the X-descent) as the base of the ventricles moves toward the apex during the ejection phase (fig. 44). The third wave (*v-wave*) results from the build-up of blood flowing from the veins while the A-V valves are closed during ventricular ejection. The mean pressure in the right atrium is about 2 mmHg, whereas that in the left atrium is slightly higher, about 5 mmHg. This is partly related to the thicker wall of the left ventricle, which is less distensible and requires a higher pressure to fill. Also, the left atrial wall is less distensible than the right. If there is an atrial septal defect, which way would the blood tend to flow?

During ventricular systole, the papillary muscles contract and, being attached to the valve leaflets through the chordae tendineae, they pull the leaflets together and prevent their eversion. They are said to contract isometrically (length unchanged) during ventricular ejection. Rupture of the papillary muscles or the chordae tendineae, due to infarction or disease, may result in acute mitral or tricuspid regurgitation. Acute mitral regurgitation may cause heart failure with acute pulmonary edema (fluid in the alveoli).

The duration of the cardiac cycle varies inversely with heart rate. If the heart beats at the rate of 75/min, the cycle equals:

$$\frac{60 \text{ sec}}{75 \text{ beats/min}} = 0.8 \text{ sec}$$

At 200 beats/min, the cycle equals:

$$\frac{60}{200} = 0.3 \text{ sec.}$$

Normally, at rest systole is shorter than diastole. At rapid rates, diastole may become shorter than systole; although both shorten, diastole shortens more.

Ventricular Blood Volume Changes

During the isovolumic contraction phase, there is no change in ventricular blood volume. The volume of blood in the ventricular cavity begins to decrease at the beginning of the ejection phase and reaches its lowest value at the end of ejection (end-systolic volume), remaining at this low point until the end of the isovolumic relaxation phase. At this time the A-V valves open and blood flows in, increasing the volume until the end of diastasis. With atrial systole, an additional volume is added to ventricular blood (end-diastolic volume) (fig. 46).

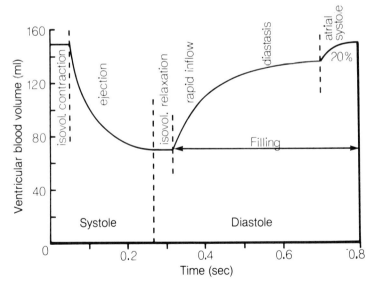

Fig. 46. Ventricular volume changes during the cardiac cycle in a resting person (75 beats/min).

Relation of ECG to Mechanical Events

Electrical changes in heart muscle precede their contraction by a short interval (fig. 47). The P-wave begins slightly before the right atrial a-wave rises. The QRS begins shortly before the two ventricular pressures rise. The end of the T-wave occurs at about the end of ventricular ejection. The ventricles are absolutely refractory until about the beginning of the T-wave. If a stimulus is applied at a "critical" moment of the T-wave, ventricular fibrillation can arise. This period is called the *vulnerable period*. If the stimulus is applied later, a premature ventricular contraction (PVC) occurs.

Note that the amplitude of the ECG gives no information about the force of myocardial contraction.

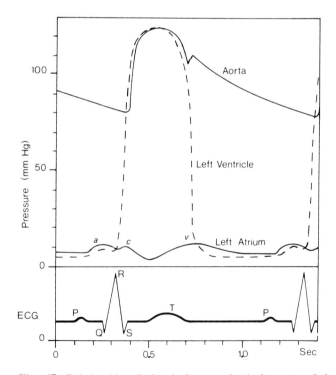

Fig. 47. Relationship of electrical to mechanical events of the cardiac cycle. [Modified from Burton, A.C.: Physiology and Biophysics of the Circulation; 2nd ed. (Year Book Medical Publishers, Chicago 1972).]

References

Burton, A.C.: Physiology and Biophysics of the Circulation; 2nd ed.; pp. 139–148 (Year Book Medical Publishers, Chicago 1972).

Noble, M.I.M.: The contribution of blood momentum to left ventricular ejection in the dog. Circ. Res. *23*:663–670 (1968).

Wiggers, C.J.: Physiology in Health and Disease; 5th ed. pp. 644–658 (Lea and Febiger, Philadelphia 1949).

7 Heart Sounds and Murmurs

The mechanical activity of the myocardium creates energy gradients in the blood (chiefly in the form of pressure) causing the closure or opening of valves depending upon the direction of the gradient and the configuration of the valve leaflets. Opening of *normal* valves causes no sounds (in valvular stenosis, opening may cause sound, eg, opening "snap" of mitral stenosis) but closure creates tension in the valves and causes vibration of the valves, heart muscle, arterial walls and of the blood. Such vibrations are transmitted to the chest wall through the intervening structures, eg, the lungs. Lung tissue, containing air, is easily compressible and dampens the vibrations. Also layers of fat in the chest wall tend to attenuate the intensity of vibrations. Some distortion is unavoidable during the transmission from the heart to the chest wall.

The frequency of vibrations on the *surface of the chest* arising from cardiac activity (in health and in disease) varies from 1 to 1000 Hz (Hertz or cycles per second). The human ear cannot hear below 20 Hz. Besides, its threshold of hearing (audibility) varies markedly with the frequency of the sound. (Refer to the threshold of hearing or audibility curve in figure 48 and note the log scale, decibels, of sound intensity.) The ear is very insensitive to low frequencies, but is most sensitive to frequencies around 2000 or 3000 Hz. Chest vibrations below 40 Hz have an intensity below the threshold of hearing, as do frequencies above 500 Hz. Hence, the frequency range of *audible* heart sounds, including murmurs, lies between 40 and 500 Hz (fig. 48).

Heart sounds may be heard by applying the ear directly to the chest or with a stethoscope (auscultation). Or, they may be recorded electrically with a *phonocardiograph*. The phonocardiograph attempts to filter out the very low frequencies electronically to imitate the audibility curve and the characteristics of the stethoscope, which is a *poor transmitter* of low frequencies. We can never be sure that such records actually represent what is *perceived* by the brain. Nevertheless, such records are

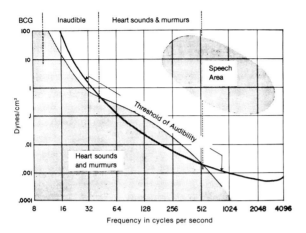

Fig. 48. The average threshold of hearing at different sound frequencies compared with the intensity of heart sounds and murmurs as heard with a stethoscope. Note that heart sound frequencies below about 40 Hz and above 500 Hz do not have sufficient intensity to be heard. Furthermore, the threshold of audibility (hearing) curve exhibits considerable variation from individual to individual, causing differences of opinion among physicians in auscultating the heart. [Reproduced, with permission, from Butterworth, J.S.; Chassin, M.R.; McGrath, R.; Reppert, E.H.: Cardiac Auscultation Including Audiovisual Principles; 2nd ed. (Grune and Stratton, New York 1960).]

useful because one can establish norms and detect abnormalities in disease. Recently, the detecting microphone has been placed at the tip of a cardiac catheter and records taken from various sites in the heart and blood vessels—intracardiac phonocardiography.

In normal persons, at least two heart sounds are heard by auscultation. Sometimes a third sound is heard. The latter is more easily recorded with a phonocardiograph. A fourth sound may be recorded (but is rarely heard).

The *first sound* (S_1) is a collection of various frequencies, ranging from 30 to 110 Hz by phonocardiography. It starts with the onset of *ventricular systole* and is associated with the closure of the A-V valves (mitral and tricuspid) (fig. 49). Excision of these valves almost completely abolishes the first sound. However, the vibrating structures responsible for the first sound include the ventricular muscle mass. Muscle contraction per se (eg, isolated muscle) causes faint vibrations, but the major mechanism of muscle vibration is believed to be the spread

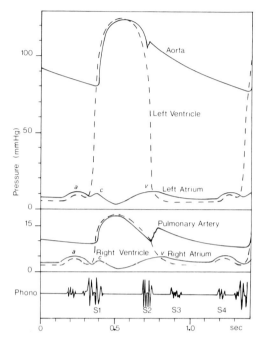

Fig. 49. The relation of heart sounds to the pressures of the chamber of the right and left heart and their respective arteries. [Modified from Burton, A.C.: Physiology and Biophysics of the Circulation; 2nd ed. (Year Book Medical Publishers, Chicago 1972).]

of valve vibration to the ventricular muscle, either directly from the valve ring or through ventricular blood. Probably both are involved. In man, the mitral valve closes *slightly before* the tricuspid, but the vibrations of the two valves usually fuse into one sound (fig. 50). In addition to these mechanisms, it is believed that ejection of blood into the large arteries with turbulent flow causes vibrations in the arterial walls, which contribute to the first sound (fig. 49). The evidence for this is not conclusive (very low frequency of vibration).

The *second sound* (S_2) is caused by closure of the aortic and pulmonic valves. It marks the beginning of ventricular diastole. In man, the aortic valve closes slightly before the pulmonic. The reason for this is unclear. The vibrating structures are the valves, the arterial walls, and probably also the blood in the large arteries. The frequency of the second sound is somewhat greater than that of the first; hence it has a

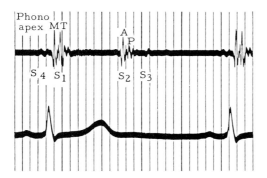

Fig. 50. Recording of heart sounds at the apex relative to the ECG. S_1 is the first heart sound; S_2 is the second; S_3, the third; S_4, the fourth. S_3 and S_4 are of low amplitude and are not heard ordinarily. M and T refer to the mitral and tricuspid components of the first heart sound; A and P are the aortic and pulmonic components of the second sound. [Reproduced, with permission, from Friedberg, C.K.: Diseases of the Heart; 3rd ed. (W.B. Saunders Co., Philadelphia 1966).]

higher pitch. The reason is that the semilunar valves and the arterial walls are under greater tension and have a smaller mass than the A-V valves and ventricular walls when these valves close. Also, the mass of blood in large arteries is less than that in the ventricles. Both of these factors cause a higher frequency of vibration. The second sound is shorter in duration (0.11 sec vs 0.14 sec) (fig. 49) because the smaller mass is more readily damped.

The opening and closing sequence of the cardiac valves of man may be summarized as follows:

Close	Open	Close	Open
MT	PA	AP	MT

where M = mitral; T = tricuspid; P = pulmonic; and A = aortic.

The *third heart sound* (S_3) may be heard in children and young persons. More frequently, it can be recorded. It occurs during diastole when blood from the atria rushes rapidly into the ventricles, tensing the chordae tendineae and the A-V ring at the end of rapid filling (especially if the ventricles are dilated or there is a large inflow) (fig. 49). The third sound is not heard readily because its frequency is very low (about 30 Hz) and our hearing threshold is high for such frequencies (lowest frequency that is intense enough to be heard is said to be 40 Hz).

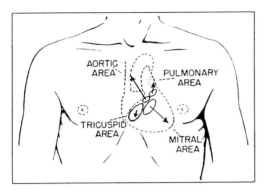

Fig. 51. Auscultation areas of the heart. Although the heart sounds are heard over the entire precordium, the vibrations from the four valves are of maximal intensity at those areas of the chest indicated by the arrows in the diagram. Refer to the text for the exact location of these areas. [Reproduced, with permission, from Rushmer, R.F.: Cardiovascular Dynamics; 3rd ed. (W.B. Saunders Co., Philadelphia 1970).]

A *fourth heart sound* (S_4) may be recorded with the phonocardiograph. It is caused by systole of the two atria and the flow of additional blood into the ventricles (during *a-wave*). Its frequency is about 20 Hz and normally is rarely heard with the stethoscope.

The loudness of heart sounds heard on the chest depends on the *site* of auscultation and the nature of intervening structures between vibrating structures and the site. Other things being equal, it varies with the *rate of change of the pressure differences* across the valves (dp/dt). Strong ventricular contractions increase the intensity of the first sound and vice versa. High end-systolic pressure in arteries increases the intensity of the second sound and vice versa (eg, pulmonary or systemic hypertension).

Sounds emitted from the four valves have their greatest intensities at four different areas on the surface of the chest. These are known as the *auscultatory areas*. These areas do not correspond to the surface anatomy of the valves. The aortic area is situated over the second right intercostal space near the sternum. The pulmonic area is over the second or third left intercostal space near the sternum. The mitral area is over the apex in the 5th left intercostal space and the tricuspid area is over the lower end of the sternum near the xyphoid (fig. 51). The physician listens to the heart sounds at these areas and notes existing abnormalities. The intensity of sounds, the rhythm, and other abnormalities are noted in each auscultation area. Rarely, the first or the second sound is split into two sounds in normal individuals. If splitting occurs normally, usually it is the second sound over the *pulmonic* area toward the end of inspiration. The reason is that inspiration decreases the intrapleural pressure; this improves filling of the right ventricle more than it does that of the left ventricle due to the difference in compliance of the two ventricles; consequently, the right ventricular stroke volume becomes slightly greater than that of the left; this prolongs the ejection of the right ventricle and delays the closure of the pulmonary valve. Another factor may be the decrease in pulmonary vascular resistance during inspiration, which would tend to prolong right ventricular ejection.

Murmurs

Valvular disease or cardiac defects that cause *turbulent* flow of blood in or near the heart result in abnormal sounds called murmurs. Sometimes murmurs occur when the heart is structurally normal, but there is an increase in the velocity of blood flow, eg, increased cardiac output.

When a murmur begins with or after the first sound and ends at or before the second sound, it is described as a *systolic murmur*. If it begins with or after the second sound and ends before the first sound, it is known as a *diastolic murmur*. If the murmur occurs throughout the cardiac cycle, it is designated as *continuous*.

Causes of Turbulence

Experimentally, turbulence occurs in blood flowing through a straight circular tube when Reynolds' number exceeds the critical value of 1000.

$$\text{Reynolds' number} \atop \text{(dimensionless)} = \frac{r \, v \, \rho}{\eta}$$

where r = internal radius of tube (cm); v = mean cross-sectional velocity (cm/sec); ρ = density (g/cm^3); η = viscosity (poise).
One poise = 1 dyne acting along 1 cm^2 for 1 second. (NB: Some authors use the diameter instead of the radius of the tube; then the critical number for blood is 2000).

From the formula it is seen that vessels with large diameters and high velocity of flow tend to have turbulence. Reynolds' number for aortic or pulmonary artery flow, which occurs only during ejection, may be calculated in man as follows:

$$N_R = \frac{1 \times 100 \times 1.06}{0.05} = \frac{106}{0.05} = 2120.$$

Therefore, blood flow through the orifices of the aorta and the pulmonary artery is normally somewhat turbulent and contributes to the first heart sound.

Turbulence can also occur from other sources. As stated above, Reynolds' number is based on studies in straight circular tubes. However, certain geometric configurations favor turbulence. The most common is the turbulence produced by a rapid flow (jet) through a narrow orifice (eg, valvular stenosis) or obstruction in a blood vessel. Additional vibrations tend to occur if the jet hits a wall such as that in a patent ductus arteriosus. In the heart, another major cause of turbulence and murmur is the occurrence of blood flow from opposing directions, as regurgitation through a valve.

Functional Murmurs

When cardiac output increases from any cause, the velocity of flow in the aorta and the pulmonary artery increases (velocity = flow/cross-sectional area). Murmurs may occur even when the valves are normal and there are no congenital defects in and around the heart. Examples of such conditions include: severe anemia, high fever, hyperthyroidism, large arteriovenous fistula, pregnancy, vigorous exercise, and so forth. These murmurs are described as *functional* or *hemic*. Are functional murmurs systolic or diastolic?

Organic Murmurs

Excluding congenital defects of the heart and adjacent vessels, organic murmurs are caused by valvular disease. *In stenosis* or narrowing of the valves, the resting cardiac output is usually maintained normal. This will increase the velocity of flow through the valve orifice (v = flow/cross-sectional area). Assuming the valvular orifice is circular there will be an increase in the Reynolds' number because velocity varies inversely with the square of radius (πr^2) when flow (stroke volume) is constant and the resulting increase in velocity outweighs the effect of a decrease in radius due to stenosis. Hence, marked turbulence and murmur occur.

In aortic or pulmonary valve stenosis, a systolic murmur occurs, the intensity of which is greatest at midsystole (described as diamond shaped) (fig. 52). The respective ventricle is hypertrophied and a large pressure gradient exists during ejection between the ventricle and its artery.

In mitral stenosis, the mean left atrial pressure is elevated (15–20 mmHg or more) and the atrial muscle is hypertrophied. There is a *dia-*

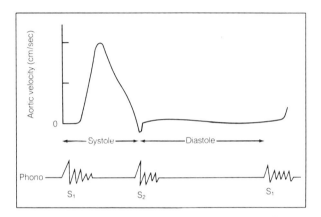

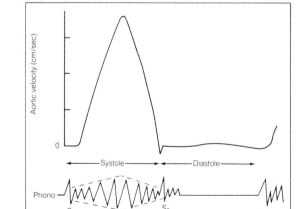

Fig. 52. Diagram illustrating the characteristic murmur that develops in stenosis of the aortic valve (b), emphasizing the importance of increased blood velocity through the narrow aortic orifice and the occurrence of turbulence. The normal aortic blood velocity and heart sounds are shown in (a) for comparison.

stolic murmur (or rumble) during the three filling phases. The murmur tends to get louder or is accentuated in late diastole or presystole, probably due to the flow caused by atrial systole (fig. 53). The opening of the stenosed mitral valve causes a short popping sound—*opening snap.*

When cardiac valves do not close properly, blood *regurgitates* into the chamber it came from. This causes turbulence in that chamber because blood enters from two opposing or different directions.

Aortic regurgitation causes early *diastolic murmur*, which diminishes in intensity as the pressure gradient between the aorta and the left ventricle diminishes during diastole (decrescendo murmur) (fig. 54). In advanced cases there may also be a short midsystolic murmur because

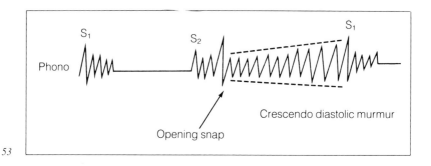

53

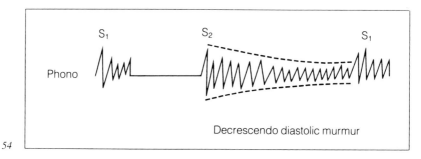

54

Fig. 53. Mitral stenosis. The murmur of mitral stenosis occurs when the valve opens during the three filling phases of diastole. The intensity of the murmur tends to increase (crescendo) when the hypertrophied left atrium pumps blood into the left ventricle.

Fig. 54. Aortic regurgitation. The improper closure of the aortic valve permits the return of some blood from the aorta to the left ventricle during diastole. The intensity of the diastolic murmur tends to decrease (decrescendo) as the pressure gradient and flow between the aorta and the left ventricle decrease during diastole. In advanced cases there may also be a systolic murmur.

the actual SV is markedly increased, causing high velocity and increased turbulence during ejection.

In mitral regurgitation there is a *systolic murmur* that persists till the second sound. It is described as a holosystolic or pansystolic murmur (pertaining to the entire period of systole). Obviously, left ventricular pressure is much higher than left atrial pressure and drives some blood back into the left atrium throughout the entire period of systole (fig. 55).

In patent ductus arteriosus the murmur is *continuous* (described as machinery-like). It peaks in late systole and early diastole, enveloping the second heart sound (fig. 56).

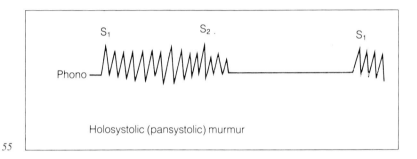

55

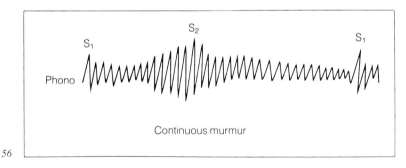

56

Fig. 55. Mitral regurgitation. Improper closure of the mitral valve causes some blood to regurgitate into the left atrium during left ventricular systole. Hence, the murmur occurs during the isovolumic and ejection phases of systole. The first heart sound is replaced by a murmur throughout the entire period of systole (pansystolic).

Fig. 56. Patent ductus arteriosus. Here blood flows continuously from the higher pressure in the aorta to the lower pressure in the pulmonary artery, causing a continuous murmur throughout the cardiac cycle. The murmur waxes and wanes more or less with the pressure gradient and flow between the two, and is described as "machinery-like."

Note that there is a systolic murmur in both aortic stenosis and mitral regurgitation and a diastolic murmur in both aortic regurgitation and mitral stenosis. One may wonder how the physician differentiates each. Several criteria are used. In the first place, the auscultatory area where each murmur is heard best (loudest) is different (aortic area *vs* mitral area). Secondly, the murmurs have different characteristics with regard to changes in intensity. Thirdly, murmurs from each valve radiate in a certain direction. Besides these characteristics of murmurs, there are other circulatory signs and symptoms that the physician utilizes to arrive at a correct diagnosis.

References

Coulter, N.A., Jr.; Pappenheimer, J.R.: Development of turbulence in flowing blood. Am. J. Physiol. *159:*401-408 (1949).

Rushmer, R.F.: Cardiovascular Dynamics; 4th ed.; pp. 411-445 (W.B. Saunders, Philadelphia 1976).

Stein, P.D.: A Physical and Physiological Basis for the Interpretation of Cardiac Auscultation. (Futura Publishing Co., New York 1981).

8 Neural Control of the Heart

The completely denervated heart (eg, transplanted), loses a significant degree of cardiac regulatory function in terms of rapid changes in the frequency of the heart beat and in the force of contraction to meet the circulatory demands of various types of stress (eg, muscular exercise). However, such hearts can deal more or less adequately with the mild or moderate stresses of sedentary life.

The frequency of contraction of the transplanted human heart is around 108 beats/min at rest but it can go up to about 125/min during exercise. Thus, the cardiac nerves, which are derived from the vagi and the sympathetics, play an important role in regulating the heart beat during various types of stress but are not absolutely essential for life.

In adult persons, the normal heart rate at rest varies between 50 and 100 beats/min depending on age, sex, intake of food, psychic condition, posture, body temperature, athletic training, and so forth. In man, a resting heart rate below 60/min is described as *bradycardia* and a rate above 100/min is designated as *tachycardia*. These limits are arbitrary and figures differ among different authors.

Influence of Vagi on the Heart

The cardiac fibers of the vagi in the neck are preganglionic fibers originating in the medulla (dorsal motor nucleus of Xth nerve and possibly the nucleus ambiguus) that synapse with ganglia located in the atrial walls near the S-A node, the A-V node, and other parts of the right and the left atrial musculature. The postganglionic fibers are short and innervate the S-A pacemaker cells, the muscle fibers of the atria, cells of the A-V node and a few fibers reach the bundle of His and its two branches. With regard to the innervation of ventricular muscle fibers, the classic teaching is that there is no vagal supply, but recent physiologic studies in dogs indicate that there is some vagal inhibitory influence. Whether

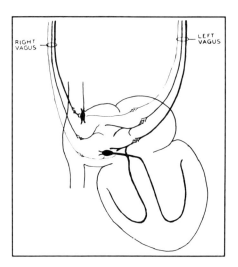

Fig. 57. Distribution of the efferent fibers of the vagi in the mammalian heart. (Relative number of fibers are indicated by the thickness of lines.)

or not this finding applies to man is uncertain. However, the coronary vessels in the ventricular walls undoubtedly receive vagal nerve supply.

The right vagus goes chiefly (but not exclusively) to the S-A node and the left vagus goes chiefly to the A-V node (fig. 57).

The actions of the vagi on the heart may be summarized as follows:

1. *Decrease or stop the autorhythmicity of the S-A pacemaker cells.* This is achieved by reducing the slope of pacemaker potential (diastolic depolarization) or by abolishing it entirely (hyperpolarization) depending on the degree of vagal stimulation (fig. 58). This action is described as a *negative chronotropic* effect.

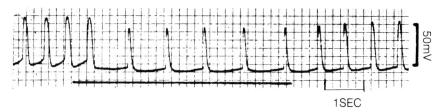

1 SEC

Fig. 58. Effect of vagal stimulation on S-A pacemaker potentials of the isolated rabbit atrium. Horizontal line below the tracing is the period of vagal stimulation. Temperature = 30°C. [Reproduced, with permission, from Toda, N.; West, T.C.: Changes in sino-atrial node transmembrane potentials on vagal stimulation of the isolated rabbit atrium. Nature *205:* 808 (1965).]

The vagi act on pacemaker cells by liberating packages of acetyl-choline (ACh) from vesicles in the postganglionic nerve endings (cholinergic nerves). The ACh acts rapidly on the pacemaker cell membrane causing its permeability to K^+ to *increase* during the period of slow diastolic depolarization. Thus, K^+ diffuses out of the pacemaker cells down its concentration gradient faster than otherwise, leaving the inside of the cell more negative. This will *reduce* the slope of pacemaker potentials and therefore reduce the frequency of firing (slow heart rate). Excessive amounts of ACh from strong stimulation of vagi can cause hyperpolarization of the membrane (approaching K^+ equilibrium potential) resulting in cessation of spontaneous discharge and cardiac arrest. Recently, it has been suggested that ACh might cause these effects by increasing the cyclic GMP in the cell membrane.

The action of ACh is short because acetylcholine is rapidly hydrolyzed by acetylcholine esterase. It is then resynthesized and stored in the vesicles of the nerve endings (for details, see biochemistry texts).

2. *Decrease the force of atrial contraction.* This is described as a negative *inotropic* effect. Recent work shows that there is also a *slight* negative inotropic action on dog ventricles.

3. *Reduce the duration of action potentials in atrial muscle.* This is achieved by increasing the rate of repolarization where the plateau almost disappears (fig. 59). As a result, vagi *shorten* the refractory period of atrial muscle. The ionic bases for these actions have not been established.

Whether or not there is a close relationship between the shortening of action potentials and the negative inotropic effect is uncertain (elec-

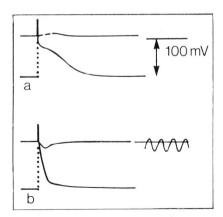

Fig. 59. Effect of A-Ch (1:2,000,-000) on transmembrane action potential of isolated dog atrial muscle. (a) = control, (b) = acetylcholine (A-Ch). Voltage calibration in (a) = 100 mV; time calibration in (b) = 20 Hz/sec. [Redrawn, with permission, from Hoffman, B.F.; Cranefield, P.F.: Electrophysiology of the Heart (McGraw-Hill Book Co., New York 1960).]

tromechanical coupling?). This is an interesting concept but in some situations it does not hold (eg, epinephrine shortens atrial action potential but increases the force of atrial contraction!).

4. *Reduce the velocity of conduction in atria, A-V node and bundle of His*. This is described as a negative *dromotropic* effect. In the ECG, one would note an increase in P-R interval or P-R segment. Experimentally, strong stimulation of the left vagus in the dog can completely block conduction in the A-V node and stop the ventricles for a short period only. Subsequently, ventricles start their own (idioventricular) rhythm from a pacemaker in some part of the Purkinje system (awakening of dormant rhythms).

The stimulation of the right or the left vagus nerve has slightly different effects on the heart depending on the differences in the anatomic distribution of the two nerves.

Chronic vagal denervation of the heart in an animal causes a distinct cardioacceleration, which leads to the conclusion that normally the vagi are continually inhibiting the heart (described as vagal tone). Full doses of atropine completely block the action of the postganglionic cholinergic fibers (described as muscarinic effect) and in man increase the heart rate by about 50 beats/min (eg, 80 beats → 130 beats/min). Therefore, in an adult at rest, vagal "tone" inhibits the rate of the cardiac pacemaker by about 50 beats/min.

The origin of vagal tone on the heart may be traced *chiefly* to afferent impulses from the carotid sinus and the aortic arch pressoreceptors. Increased afferent impulses from these receptors cause greater tone of the vagi and vice versa (see details in Ch. 22).

Newborn infants have little cardiac vagal tone. With growth, vagal tone develops. The mechanism of this development is not known. One might suspect that perhaps it is related to the rise of blood pressure with age, which induces the flow of impulses from the pressoreceptors. Of course, it may also be a genetic development.

Influence of Sympathetics on the Heart

The chief center of the cardiac sympathetic nerves is in the medulla, but the neurons are not grouped together like those of the vagi which can be identified under the microscope. This center controls the sympathetic *preganglionic* neurons located in the lateral horns of the spinal cord from T_1–T_5 (fig. 60). Most preganglionic fibers synapse in the ganglia of the sympathetic chains (T_1–T_5), but some do so with neurons in the middle and superior cervical ganglia; a few synapse in the cardiac plexus near the heart. Postganglionic fibers are relatively long, running in the superior, middle and inferior cardiac nerves. Others come from T_2–T_5 ganglia in the sympathetic chain and reach the cells of the S-A node, atrial muscle, A-V node, bundle of His and its branches, Purkinje system. In addition, *a large number innervate the ventricular muscle fibers*. The ventricular distribution is important functionally, because these nerves augment ventricular "contractility" and under certain conditions, they cause dormant pacemakers in the Purkinje fibers to wake up, resulting in ventricular arrhythmias.

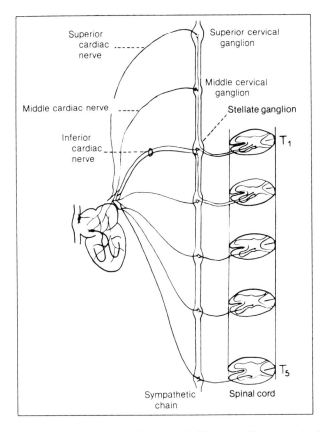

Fig. 60. Origin and distribution of efferent cardiac sympathetic nerves (man).

The actions of cardiac sympathetics may be summarized as follows:
1. *Increase the frequency of the S-A pacemaker.* This is described as a positive chronotropic effect. It is achieved by increasing the slope of pacemaker potentials (fig. 61). The ionic basis for this is not established. According to Noble the increased slope is due to increased sodium conductance (gNa^+) of the membrane during slow diastolic depolarization. In other words, sodium enters more rapidly than normal and the membrane potential reaches the threshold more rapidly. Others do not agree with this concept.

The sympathetic postganglionic transmitter is norepinephrine (NE). In isolated Purkinje fibers that are quiescent, large doses of NE induce pacemaker potentials and firing. In the intact animal, large intravenous doses of catecholamines tend to *slow the heart reflexly* (baroreceptor reflexes from rise of blood pressure outweigh the direct action on pacemaker) and may cause ectopic rhythm in ventricles (awakening the dormant pacemaker activity of Purkinje tissue).

NE is said to act on *β_1-receptors* in cardiac cell membrane causing activation of *adenylyl cyclase* which converts ATP into cyclic AMP. Cyclic AMP is believed to cause the various cellular actions and is described as a *second messenger*. c-AMP is then reconverted into ATP.

NE acts on the heart relatively more slowly than ACh and its action lasts longer. It is oxidized by two enzymes found in cardiac cells: (a)

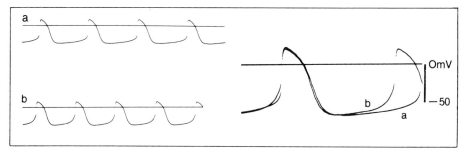

Fig. 61. Effect of sympathetic nerve stimulation on pacemaker potentials of isolated rabbit S-A pacemaker cell discharging spontaneously. (a) = control; (b) = sympathetic stimulation. (*Right*) Superimposed (a) and (b). [Reproduced, with permission, from Toda, N.; Shimamoto, K.: The influence of sympathetic stimulation on transmembrane potentials in the S-A node. J. Pharmacol. Exp. Ther. *159:* 298 (1968).]

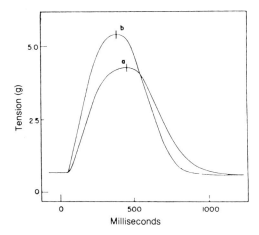

Fig. 62. Isolated cat papillary muscle stimulated 30 times per minute at 23°C. Superimposed isometric contractions. (a) Control; (b) norepinephrine (0.05 μg/ml). [Reproduced, with permission, from Sonnenblick, E.H.: Force-velocity relations in mammalian heart muscle. Am. J. Physiol. *202:* 931 (1962).]

monoamine oxidase (MAO) found chiefly in mitochondria and (b) catechol-O-methyl transferase (COMT) found largely in the cytosol.

2. *Increase the peak contractile tension of atrial and ventricular muscles and shorten the time to peak tension* (fig. 62). This is described as a positive inotropic effect.

   The force-velocity curve is shifted upward and to the right (increased P_0 and V_{max}). The increased V_{max} is taken to indicate an increase in "contractile" or "inotropic" state of the muscle.

3. *Shorten the duration of action potential in atrial and ventricular muscle* (fig. 63). Therefore, sympathetics shorten the refractory period of atria and ventricles. Note that the shortening of the action potential is associated with a positive inotropic effect, contrary to that of vagus.

4. *Velocity of conduction is increased in all parts of the heart.* Positive dromotropic effect. If the ECG is recorded, stimulation of sympathetics would cause shortening of the P-R and the QT intervals.

   Note: In the intact heart, cardioacceleration by *electrical pacing* shortens the duration of action potential, increases conduction velocity and has complex effects on contractility. Thus, heart frequency per se has significant effects on the heart, the mechanism of which is not clear.

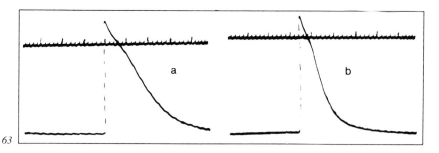

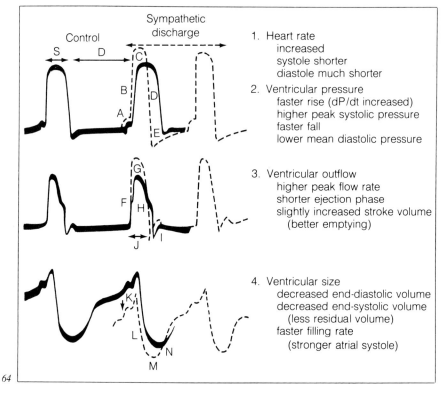

Fig. 63. Effect of epinephrine on transmembrane action potential of a dog atrial fiber. (a) Control; (b) epinephrine. Time marks in 10 and 50 msec intervals. [Reproduced, with permission, from Brooks, C.M.; Hoffman, B.F.; Suckling, E.E.; Orias, O.: Excitability of the Heart. (Grune and Stratton, New York 1955).]

Fig. 64. Influence of sympathetic nerves on the heart. Note that most of the changes are related to the acceleration of rhythmic and contractile processes of the myocardium (positive chronotropic, dromotropic and inotropic effects). [Modified, with permission, from Rushmer, R.F.: Cardiovascular Dynamics; 4th ed. (W.B. Saunders Co., Philadelphia 1976).]

The effects of stimulating cardiac sympathetics in an intact animal are summarized in figure 64.

The tone of the sympathetic nerves to the heart may be studied by sympathectomy or by the use of β_1-receptor blocking drugs, eg, propranolol, practolol and so on). Full doses of propranolol in man slow the heart by about 25 beats/min. Hence, sympathetic tone under resting conditions is less than vagal, but it plays an important role in augmenting both the frequency and the *force* of myocardial contraction. Total block of both cardiac sympathetics and parasympathetics (atropine + propranolol) in adult human subjects increases the heart rate to about 105 beats/min. This discloses the *intrinsic rate of the pacemaker* (fig. 65) (agrees with rate of transplanted hearts).

Athletic training slows the resting heart rate (well-known bradycardia of athletes), but cardiac output and arterial pressure remain normal. Formerly, it was thought that the bradycardia is due to an increase in vagal tone. Recent studies do not support this view and suggest two

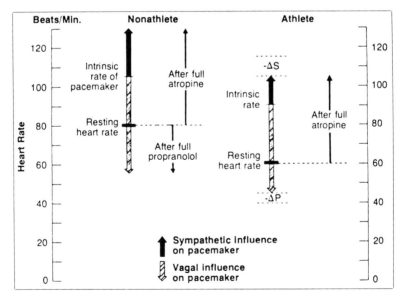

Fig. 65. Diagram of the resting bradycardia produced by exercise training. The two factors that seem to be involved are (a) slowing of intrinsic rate of cardiac pacemaker, and (b) reduction of sympathetic influence on pacemaker with slight decrease or no change in vagal influence. [Reproduced from Badeer, H.S.: Axioms on athlete's heart syndrome. Hosp. Med. *13/4:* 70 (1977).]

factors: (a) slowing of the "intrinsic" rate of the S-A node from 105 to 90/min by some unknown mechanism, and (b) decreased influence of cardiac sympathetics with no change or a slight decrease in vagal influence on the pacemaker (fig. 65). The mechanism bringing about these changes is unknown.

The chief medullary centers controlling the heart are referred to as the *cardioinhibitor* and *cardioaccelerator* centers. These centers are under the influence of higher centers in the hypothalamus and other parts of the brain. The neural pathways from these highest centers are not well worked out.

Vagal tone to the heart may vary with the respiratory cycle. Inflation of the lungs during inspiration leads to decreased cardiac vagal tone, thereby accelerating the heart and vice versa. The afferent nerves from the lungs travel in the vagi and are sensitive to stretch. In children, the variations in heart rate with respiration may be very marked, a condition called *respiratory arrhythmia*.

Afferent Nerves From the Heart

Afferent nerve fibers from the heart travel in both the vagi and the sympathetics:
1. Vagal afferents. These are chiefly involved in reflexes to be discussed later (stretch receptors in atria regulating blood volume, and probably there are also receptors in the ventricles).
2. Sympathetic afferents (cell bodies are in the dorsal root ganglia). These fibers mediate *pain* sensation from the region of the myocardium, eg, angina pectoris or myocardial infarction. Such pain receptors are insensitive to the cutting of heart muscle.

There are no touch receptors in the heart, as indicated by the fact that catheters placed inside the heart are not felt by the unanesthetized subject.

References

Badeer, H.S.: Resting bradycardia of exercise training: a concept based on currently available data; in Roy and Rona: Recent Advances in Studies on Cardiac Structure and Metabolism; vol. 10; The Metabolism of Contraction; pp. 553–560 (University Park Press, Baltimore 1975).

Randall, W.C.: Neural Regulation of the Heart (Oxford University Press, New York 1977).

9 Cardiac Output: Measurement and Determinants

Since the heart is a double pump with two vascular circuits placed in series (fig. 66), the amount of blood pumped out by the right ventricle must be equal to that of the left over a period of time. Otherwise,

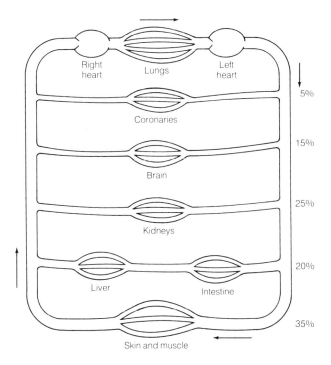

Fig. 66. Schematic diagram of the circulation showing the relation between the systemic and the pulmonary vascular circuits. The percentages indicate the distribution of left ventricular output to the various major parallel circuits in man under resting conditions. [Redrawn from Rein, H.; Schneider, M.: Physiologie des Menschen; 12th ed. (Springer-Verlag, Heidelberg 1956).]

blood would pile up in one of the circuits, eg, 1 ml difference per beat if kept up *continuously* for 1 hour would collect practically all the blood in one of the circuits (80 beats/min = 80 ml/min; 80×60 min = 4800 ml/hr. However, under transient or *unsteady* states, there may be slight inequality in the output of the two ventricles.

Definition

Cardiac output (CO) is the volume of blood pumped out by *each* ventricle *per minute* (not the sum of the volumes pumped out by the two ventricles). The reason for defining it for *one ventricle alone* is twofold:
1. It is the meaningful flow for the hemodynamics of each circuit, and
2. It is the meaningful blood flow for the function of each circuit.
The unit of time is conventional. The term cardiac or heart is, therefore, misleading and strictly speaking is a misnomer. A more appropriate term would be ventricular output. However, the term cardiac is so well established that it would be inadvisable to change it.
Under steady states:

Left ventricular output = right ventricular output (99%) + bronchial circulation (1%).

The bronchial vessels (derived from the systemic circuit) drain into the pulmonary capillaries and veins and thereby by-pass the right ventricle.
By definition:

Cardiac output = stroke volume × heart rate/min.

In order to equalize the right and left ventricular outputs, the two ventricles are intimately attached to each other and beat at the *same rate* from the discharge of one pacemaker (ingenious design!). In addition, there are very effective intrinsic and extrinsic regulatory mechanisms that adjust and equalize the stroke volume of one ventricle with that of the other on a long-term basis.

Measurement of Cardiac Output in Man

The first two methods that will be considered are invasive and are based on the *dilution principle*. The simplest case of this principle may be illustrated in the indirect measurement of the volume of a stationary liquid. A known amount of a solute is dissolved in an unknown volume of its solution and the change in concentration determined, eg, 10 grams sugar are added to x-volume of 1% solution of sugar in a container; after dissolving *completely*, sugar concentration rises to 3%; what is the volume of the solution?

Answer:

$$\frac{10 \text{ g}}{\dfrac{3 \text{ g}}{100 \text{ ml}} - \dfrac{1 \text{ g}}{100 \text{ ml}}} = \frac{10 \text{ g}}{\dfrac{2 \text{ g}}{100 \text{ ml}}} = \frac{10 \text{ g} \times 100 \text{ ml}}{2 \text{ g}} = 500 \text{ ml}$$

Direct Fick Method. Applies the dilution principle in a flowing fluid under *steady state* conditions. If the amount of a substance which *steadily enters* or leaves a flowing stream is known and the change in its concentration resulting therefrom is determined, one can calculate the volume of fluid that is flowing at a uniform rate. This principle is applied to the uptake of O_2 by the flowing blood in the capillaries of the lungs (fig. 67).

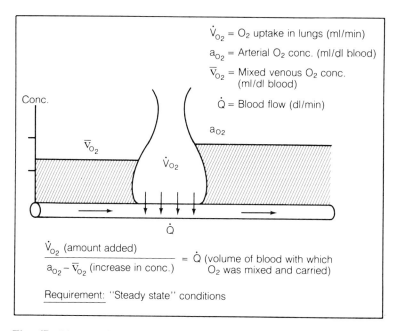

$\dot{V}_{O_2}$ = O_2 uptake in lungs (ml/min)

a_{O_2} = Arterial O_2 conc. (ml/dl blood)

$\overline{V}_{O_2}$ = Mixed venous O_2 conc. (ml/dl blood)

$\dot{Q}$ = Blood flow (dl/min)

$$\frac{\dot{V}_{O_2} \text{ (amount added)}}{a_{O_2} - \overline{V}_{O_2} \text{ (increase in conc.)}} = \dot{Q} \text{ (volume of blood with which } O_2 \text{ was mixed and carried)}$$

Requirement: "Steady state" conditions

Fig. 67. Diagram illustrating the principle of the Fick method in determining cardiac output in the intact animal.

Essential Conditions
1. Steady state: (a) mean flow must be constant (constant heart rate, stroke volume, total resistance of vessels, and so on); and (b) uptake or release of the substance must be uniform (constant rate and depth of breathing, gas concentration, body at rest, and so on).
2. Substance should *remain in the stream* and must neither be manufactured nor destroyed in the measured part of the system and must be *fully mixed* with blood.

Calculation

$$\frac{O_2 \text{ uptake in lungs/unit time } (\dot{V}_{O_2})}{\text{Increase in } O_2 \text{ concentration } (a_{O_2} - \bar{v}_{O_2})} = \text{blood flow/unit time } (\dot{Q})$$

where a_{O_2} = arterial O_2 concentration; $\bar{v}_{O_2}$ = O_2 concentration in *mixed* venous blood in pulmonary artery.

In practice, arterial blood is obtained by puncture of a peripheral artery (eg, femoral) and *mixed* venous blood is obtained from the *pulmonary artery* by a catheter (cannot use blood from peripheral veins because their O_2 content differs) and O_2 uptake in lungs by spirometry or gas analysis (fig. 68). The accuracy of the method is about ± 10%.

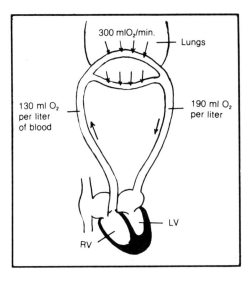

Fig. 68. Scheme illustrating Fick's principle for measuring cardiac output. RV = right ventricle; LV = left ventricle.

In a resting person lying *supine*, the average values are:

$$\frac{250 \text{ ml } O_2/\text{min}}{190 - 145 \text{ ml } O_2/\text{liter blood}} = \frac{250}{45} = 5.5 \text{ liters/min.}$$

Note: O_2 concentration is often expressed in volumes percent (vol %) or milliliters per deciliter (ml/dl) blood (1 deciliter = 100 ml).

$$\text{Mean stroke volume} = \frac{5500 \text{ ml/min}}{80 \text{ beats/min}} = 70 \text{ ml/beat (approx.).}$$

Cardiac output in resting individuals varies with body size, sex, posture, and so on. The best correlation is with body surface area. In resting adult persons CO = 3.2 liters/min per square meter of body surface. This is called the *Cardiac Index*. (The average surface area of an adult = 1.73 m².)

Can the CO_2 expired be used to calculate cardiac output? Theoretically yes, but in practice the results are much more variable because slight changes in ventilation rate ($\dot{V}$) alter the CO_2 content of arterial blood significantly. This is due to the characteristics of the blood CO_2 transport system compared with the hemoglobin oxygen dissociation curve which is flat at its upper part. Hence arterial O_2 content does not fluctuate readily with changes in $\dot{V}$ (see Respiration).

Stewart-Hamilton Method. This method applies the dilution principle when flow is steady but the indicator is added in *one shot* (not continuously) into the blood of the right heart. A known amount of indicator is *rapidly* injected into the right heart through a catheter and its *mean* arterial concentration is calculated by integrating the infinitesimal changes in concentration with respect to time at any one point. Mean arterial concentration = area under the time-concentration curve divided by time "t" which is the time for *one* circulation of the injected indicator. Figure 69 shows this in a simplified model of the heart. In the mammalian circulation, the indicator shows a second rise of concentration in the arterial blood due to recirculation through the coronary and lung circuits (fig. 70). This must be eliminated to obtain the "t" of the *first* circulation. It is done by a "semilog" plot where the concentration is expressed on a log scale (the *exponential rise and fall* of concentration becomes a *linear* relationship on a log scale and can be ex-

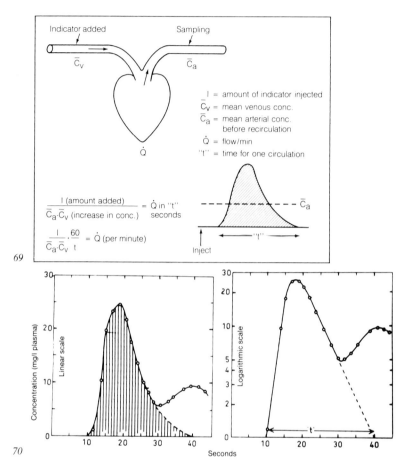

69

70

Fig. 69. Principle of the Stewart-Hamilton method. An indicator is added to venous blood near the heart and its concentration in arterial blood rises and falls exponentially (model shows one ventricle for simplicity). Blood flow or cardiac output is equal to the amount injected divided by the increase in concentration of indicator. The time for one circulation, which is less than 60 sec, is then adjusted for 1 min since output, by definition, is outflow/min.

Fig. 70. Stewart-Hamilton method of determining cardiac output. The linear scale plot on the left is the time-concentration curve of indicator (Evans blue, T-1824) in arterial plasma after a rapid intravenous injection. The second rise is due to recirculation of indicator through the coronary circuit. To eliminate the recirculation rise of concentration the semilog plot on the right is carried out. This gives a straight line rise and fall of arterial concentration whereby extrapolation of the descending limb will permit the determination of time "t" for one circulation. The time "t" on the linear plot will permit measurement of the area under the curve for one circulation from which the mean concentration can be calculated: mean concentration (C) = area under linear plot / "t".

trapolated easily to find time "t") (fig. 70). Note that the indicator must not leave the system during this period.

$$\frac{\text{Amount of indicator added}}{\text{Increase in mean arterial conc. during "t"}} = \text{flow/time "t"}$$

Flow/unit time = flow/"t" × unit time (60 sec)

Many indicators have been used for the purpose, eg, various dyes (cardiogreen), radioiodinated serum albumin (RISA), hydrogen gas, cold saline (thermodilution), and so on. Time-concentration curves are now *recorded* by appropriate means (counters, densitometers, H_2 electrode, thermistors, and so on). The curves can be integrated electronically and cardiac output obtained by digital readout. Currently, the popular method is to inject cold saline into the right atrium by means of a double lumen catheter whose tip lies in the pulmonary artery and contains a thermistor which registers the changes in temperature continuously. In the thermodilution method the problem of recirculation and errors of extrapolation are eliminated (saline warms up before recirculation occurs). A small amount of heat may be taken up from the wall of the right ventricle by the cold saline before reaching the pulmonary artery introducing a slight error.

The advantage of this method over Direct Fick is that it can be repeated several times at short intervals and the results obtained quickly with electronic integration. Therefore, one can study cardiac output in nonsteady circulatory states such as exercise, action of drugs, and so forth.

Direct Fick and Stewart-Hamilton methods have checked well with each other when carried out simultaneously (slope is about 45° or tangent is 1.0).

Echocardiography (Noninvasive Method)

High frequency sound (2.25 megahertz) is applied over the skin of the precordium between the ribs and the *reflected* sound waves from various interfaces of the chest and heart are picked up by the piezoelectric crystal that emits the sound. These waves are analyzed and recorded graphically by the so-called M (motion) mode of recording. The record looks quite complicated due to the existence of various reflecting inter-

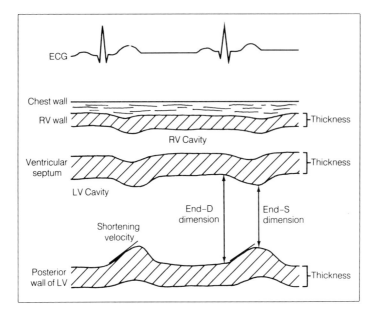

Fig. 71. Schematic drawing of M-mode echocardiogram showing the changes in left ventricular dimensions during a normal cardiac cycle (chordae tendineae are not shown).

faces and the constant dynamic state of the heart. Figure 71 is a schematic drawing.

From this tracing one can study and analyze various structures of the heart (eg, valves, ventricular wall thickness, etc.) in health and in disease and the motions they undergo during systole. If the left ventricle is assumed to be a *prolate ellipse,* ie, a cylinder whose diameter (D) is ½ the longitudinal axis (L), and the minor axis (D) is measured from the echo tracing, one can calculate the volume of the left ventricular cavity.

$$\text{Volume of LV} = \frac{\pi}{6} D_1 \cdot D_2 \cdot L$$

D_1 and D_2 are the two minor axes and L is the longitudinal axis. If D_1 and D_2 are assumed to be equal, then

$$V = \frac{\pi}{6} D^2 L = \frac{\pi}{6} D^2 \cdot 2D = \frac{\pi}{3} D^3 = D^3$$

Thus, the volume of the left ventricle is roughly equal to the 3rd power of the transverse diameter (or minor axis). If one measures D at the end of diastole (D_d) and at the end of systole (D_s), it becomes easy to calculate the stroke volume.

Stroke volume $= D_d{}^3 - D_s{}^3$

Such calculations involve many crucial assumptions and inaccuracies. The values obtained are not precise, specially when the heart is dilated. The reason for the latter is that the dilated left ventricle is more spherical and L is no more equal to 2D. For the dilated ventricle, other formulas have been proposed.

EDV $= 59$ (EDD) $- 153$
ESV $= 47$ (ESD) $- 120$
Stroke volume $=$ EDV $-$ ESV

From the stroke volume one can calculate the cardiac output by knowing the heart rate. This method is not as accurate as the preceding two and caution must be exercised in drawing conclusions.

Factors Determining Cardiac Output in the Isolated Heart

This was first studied in the isolated heart perfused with blood. The pioneer work was done by Starling and co-workers [1912–1914] on the dog heart-lung preparation which they designed (fig. 72). In brief, the heart is exposed under artificial respiration and the aortic blood is diverted through the brachiocephalic artery into an artificial circuit whose resistance to flow can be altered by pumping air around an enclosed thin rubber tube (often referred to as Starling resistance). The blood is kept warm by passing through a coil in a constant temperature water bath and is then returned to a venous reservoir which empties its blood into the superior vena cava. The pulmonary circuit is left intact to oxygenate blood and the coronary circulation maintains the heart beat. All other vessels to and from the heart are tied. The heart is denervated (vagi cut) and the body of the animal is dead from lack of circulation. Left ventricular output (minus coronary flow) can be measured by diverting the blood for 5 seconds into a measuring cylinder located beyond the Starling resistance.

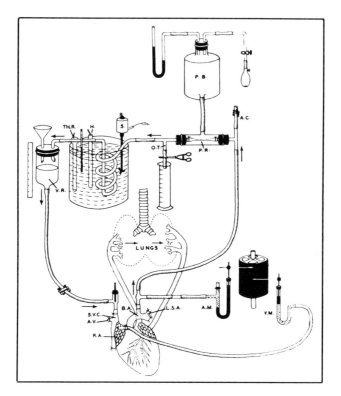

Fig. 72. Arrangement of the artificial circuit in Starling's heart-lung preparation. (The different parts are not drawn to scale, and the body of the animal is not drawn). BA = brachiocephalic artery; LSA = left subclavian artery; AM = arterial manometer; AC = air cushion; PR = peripheral resistance; PB = pressure bottle; OT = output tube; S = stirrer; H = heater; THR = thermoregulator; VR = venous reservoir; SVC = superior vena cava; AV = azygous vein; RA = right auricle; VM = venous manometer.

The advantage of this preparation is that the arterial pressure, venous inflow and the heart rate can be independently controlled. Starling and his co-workers investigated the influence of each of these factors on cardiac output (CO).

Influence of Peripheral Resistance on CO

In this study, heart rate and venous inflow are kept constant and the peripheral resistance (hence arterial pressure) is increased or decreased by pumping more or less air around the thin tube (PR). The result is shown in figure 73. The conclusion is that CO is independent of pe-

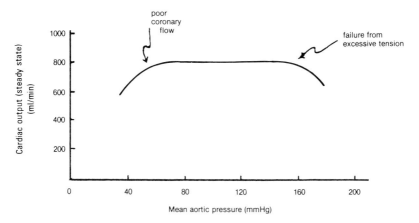

Fig. 73. Dog heart-lung preparation. Output set at about 800 ml/min. Mean aortic pressure is altered by changing the peripheral (Starling) resistance. Heart rate and venous inflow are kept constant. Note that output under steady states remains independent of aortic pressure in a range from 60 to about 160 mmHg. Below 60 mmHg, output falls from poor coronary flow and contractility; above 160 mmHg it falls from excessive pressure and tension in the wall of the ventricle. Residual blood volume in the ventricles increases as aortic pressure is increased, but the stroke volume is kept constant by the Frank-Starling mechanism up to a certain limit.

ripheral resistance or aortic pressure within a wide range of pressures (60–160 mmHg in such a preparation). If the aortic pressure is very low, the heart fails because coronary flow is inadequate; if pressure is very high, the heart fails to overcome the excessive pressure.

Influence of Venous Inflow on CO

In this study, heart rate and arterial pressure are held constant, and venous inflow altered by raising or lowering the venous reservoir. It is seen in figure 74 that CO varies directly with ($y = kx$) and is equal to the venous inflow ($k = 1$, or slope $= 45°$), within limits. Excessive inflow overstretches the heart and the muscle fails (descending limb of length-tension curve).

Influence of Heart Rate on CO

Here the venous inflow and peripheral resistance are held constant. Figure 75 shows that, within limits, changes in heart rate do not alter the CO. If the heart rate is too fast, CO drops from inadequate ventricular filling and if it is too slow, the heart is overfilled with each diastole and begins to fail from overstretching (descending portion of length-tension curve).

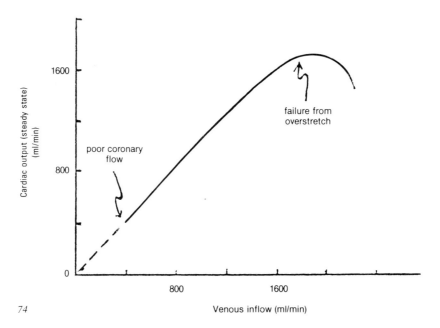

74

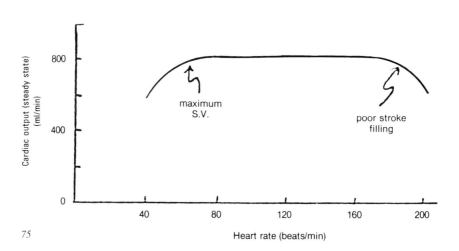

75

Conclusion

In a *denervated isolated heart* under steady states, cardiac output is determined by and is equal to the venous inflow, within certain limits. *If venous inflow is held constant,* output is independent of heart rate and peripheral resistance, within limits. Many people erroneously equate cardiac output with heart rate. This would be true *only if* heart rate would vary directly with and to the same degree as the venous inflow (slope 45°). In the *intact* organism, this is true in a number of situations but not under all circumstances. Hence, caution must be exercised. *The basic factor that determines cardiac output is venous inflow, provided the heart is not failing acutely* during which output is less than inflow for a short period. It is implicit in the foregoing conclusion that the heart is beating and pumping. We shall later consider the factors that determine venous return to the ventricle in the intact circulation and note that the driving force for venous return is the energy or pressure gradient in the venous system which is ultimately derived from the pressure developed by the ventricle. Hence, cardiac output and venous return are *interdependent variables* closely coupled in a very complex manner.

Some authors object to the distinction between venous return and cardiac output on the grounds that in a closed circuit, like the circulation, under steady states cardiac output and venous return must be and are equal. Consequently, it is said to be irrelevant whether one thinks

---

Fig. 74. Dog heart-lung preparation. Heart rate and aortic pressure are kept constant as venous inflow is reduced and increased gradually by lowering and raising the venous reservoir. Note the linear relationship between venous inflow and cardiac output in a wide range. Ventricular size increases as venous inflow increases. Stroke filling increases, leading to an increase in stroke volume by the Frank-Starling mechanism. When filling is excessive, the optimal initial length of ventricular fibers is exceeded, and the descending limb of the curve develops.

Fig. 75. Dog heart-lung preparation. With venous inflow and peripheral resistance constant, heart rate is altered by crushing the S-A node and pacing one of the atria electrically. Note that cardiac output is independent of heart rate when venous inflow is constant. At the extremes of heart rate, cardiac output falls either because the optimum fiber length is exceeded (stroke volume falls below its maximum) or ventricular filling time is too short to fill the ventricles adequately (eg, 180 beats/min or more).

of the flow around the circuit as "cardiac output" or "venous return." This is obviously true *only under steady state conditions;* not so during transient or nonsteady states in which case the distinction becomes very significant. In everyday life nonsteady states occur more frequently than not (eg, muscular activity, postural changes, etc.) and the temporary inequality of venous return and cardiac output assumes great importance in the understanding of circulatory adjustments in health and in disease. The subject of venous return will be considered in more detail in Chapter 20.

Starling analyzed the mechanism of these adjustments by measuring the volume changes of the ventricles with a "cardiometer." He found that increase in venous inflow → increases end-diastolic volume → greater "initial" fiber length → stronger contraction → increased stroke volume and output (heart rate and blood pressure constant). Similarly, there was an *increase* in end-diastolic volume but *no change* in stroke volume when arterial resistance was increased. This comes about as follows: Upon increase in peripheral resistance, left ventricular contraction is unable to pump its previous stroke volume thereby increasing the residual blood in the ventricle. Venous inflow and ventricular filling being unchanged, the end-diastolic blood volume increases causing more forceful contraction of the ventricle and improved stroke volume. This chain of events continues for several beats until the increase in end-diastolic volume is such that the augmented force of ventricular contraction is capable of restoring the stroke volume against the increased aortic pressure. From these observations Starling enunciated the "law of the heart" which states that the *force of contraction is a function of the initial length of heart muscle fibers,* up to an optimum. Starling's observations were in agreement with earlier studies of Otto Frank on the frog heart. Frank had recorded the *peak* pressures developed by the frog ventricle during *isovolumic* contractions when muscle length was progressively increased by increasing ventricular filling. Contractile "peak" pressures increased with greater filling up to a maximum and then declined (fig. 76).

These findings were in agreement with the length-tension diagram of isolated heart muscle strips or papillary muscle. Such curves came to be known as the Frank-Starling curves. Starling extrapolated the frog pressure-volume curve to the pressure-volume curve of the dog left ventricle and Katz has further extrapolated it to the human left ventricle (fig. 77).

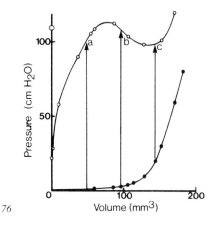

76

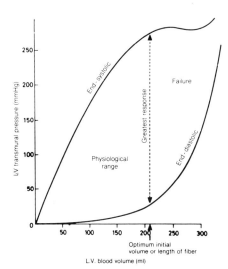

77

Fig. 76. Pressure-volume curve of the frog ventricle. The aorta is occluded and the single ventricle is filled with increasing amounts of fluid, contracting isovolumically. The end-diastolic and peak systolic pressures are recorded and plotted. Note the initial increase (ascending limb) and then decrease (descending limb) in peak developed or contractile pressure, as the volume increases. The curve is similar to the length-tension curve of isolated strips of heart muscle. [Modified from Rein, H.; Schneider, M.: Physiologie des Menschen; 12th ed. (Springer-Verlag, Heidelberg 1956).]

Fig. 77. Estimated pressure-volume relationship in the beating human left ventricle in the absence of ejection (perfused coronaries). The end-systolic curve represents the maximum pressure developed with different volumes in the left ventricle when ejection is prevented (isometric maximum). The values for man are extrapolated from experiments on the isolated frog and dog hearts.

The ultrastructural basis of Frank-Starling mechanism is believed to be the number of active or force generating crossbridges between myosin and actin filaments in the sarcomere prior to the beginning of contraction. (Refer to skeletal muscle physiology).

In short, the completely denervated heart (eg, transplanted) adjusts and equalizes its output to the venous inflow by changing *chiefly* its stroke volume on the basis of the Frank-Starling mechanism. Sarnoff called this mechanism *heterometric autoregulation*. The completely denervated heart, however, can increase its rate to a moderate extent. Possible mechanisms for this would include stretch of S-A node, circulating catecholamines from the adrenal medulla, a rise in body temperature, excess thyroid hormone, and so on.

Functions of the Pericardium

The pericardium consists mostly of collagenous tissue which is not very distensible. Consequently, the pericardium functions to prevent acute overdistension of the heart and thereby curtail the descending limb of the Frank-Starling curve. In addition, the smooth surface of the pericardium together with the small amount of liquid contained in the sac serve to reduce the friction between the contracting heart and the parietal pericardium.

Output of the Intact Innervated Heart

The question may be asked: is venous inflow the major determinant of cardiac output in the intact innervated heart? There is every reason to believe that it is so (provided the heart *is not* in *acute* failure). If this is true, what are the available mechanisms that the intact heart may utilize to equalize cardiac output and venous inflow whenever there is a change in venous inflow?

In a stroke pump (such as the heart), output can be altered in one or both of two ways: (a) varying the number of strokes, or (b) varying the volume of each stroke. Studies in animals and man have shown that, more often than not, the intact heart makes adjustments to venous inflow by changing its rate rather *than its stroke volume,* but both means may be utilized depending upon circumstances. Under physiologic conditions, venous inflow and cardiac output are increased during muscular exercise, after intake of food and drink, on exposure to heat, on changing from a standing to a lying position, on exposure to cold with shivering, in anxiety associated with increased metabolic rate, and so on. In most of these circumstances (with the exception of lying down) the heart takes care of the increase in venous inflow *primarily by increasing its rate*. This has led to the erroneous idea that cardiac output increases because of increase in heart rate. In fact, the basic factor is venous inflow and heart rate is one of the means of equalizing output to inflow. *If venous inflow per minute is constant, increased heart rate cannot increase output.* This has been demonstrated not only in the heart-lung preparation but also in intact man and animals by electrically pacing the heart under *resting* conditions. The factors that determine or change venous inflow will be discussed in Chapter 20.

Alterations in Heart Rate

Rapid changes in heart rate are accomplished by altering the tone of the vagi or of the cardiac sympathetics on the pacemaker. Such changes may be brought about reflexly through the medullary centers (eg, fall of pressure in the carotid sinuses upon standing increases heart rate) or by influences from higher centers of the brain (eg, exercise, exposure to heat, and so forth).

During muscular activity, venous return is increased and cardioacceleration occurs as a result of decreased vagal tone and increased sympathetic tone. The mechanism underlying these changes is not clear but some experiments of Rushmer [1959], suggest the involvement of hypothalamic centers. According to Rushmer [1959], mild or moderate exercise in nonathletic individuals in the *supine* position increases output by means of cardioacceleration, with little or no increase in stroke volume. In contrast, in the *erect* position where the stroke volume is less than in supine position there is a significant increase (10–30%) in stroke volume during exercise. In athletes the increase in stroke volume is more marked. In both the nonathlete and the athlete the *primary factor in increasing output is the increase in venous inflow to the heart* at the beginning of exercise before a steady state is reached.

Heart rate increases in exposures to ambient heat or cold and in emotional disturbances. It is believed that the hypothalamus and the cerebral cortex are involved.

Heart rate increases after intake of food, but the mechanism is unclear.

Besides neural control of pacemaker, there are non-neural influences that may act directly on pacemaker frequency:

1. Circulating catecholamines from adrenal medulla → accelerate pacemaker.
2. Stretch of S-A node → accelerates pacemaker by reducing the negativity of maximal diastolic potential and to a minor degree by increasing the slope of pacemaker potential.
3. Rise of pacemaker temperature (eg, fevers) → accelerates the pacemaker by increasing the slope of pacemaker potential and vice versa.
4. Excess circulating thyroid hormone causes cardioacceleration but it has a relatively long period of latency (takes hours or days to act), eg, rapid pulse of hyperthyroidism is characteristic.
5. Marked decrease in arterial P_{O_2} (hypoxemia), eg, drop from normal of 95 torr to 60 torr or increase in arterial P_{CO_2} → accelerate pacemaker rhythm.
6. Hyperkalemia (increased blood K^+) increases the frequency of the isolated S-A node and vice versa (opposite is true for Purkinje fibers). High blood levels of K^+ favor ventricular premature systoles, ventricular tachycardia and ventricular fibrillation and arrest the heart in a relaxed state.

In isolated pacemaker tissues, very marked decrease of extracellular calcium accelerates the rhythm, but in the body such low Ca^{2+} levels are incompatible with life.

In disease, toxins of bacteria or viruses may alter heart rate by acting directly on nerve centers controlling the heart or on pacemaker tissue or by changing body temperature and the circulatory state. In general, heart rate increases with fevers but in some conditions the heart rate may be slow although body temperature is high (eg, typhoid fever). Certain drugs may alter heart rate in a complex manner.

Changes in "Contractility" and Stroke Volume

Contraction of any muscle is manifested either as a change in the *external length* of the muscle or in the development of *force*, or, more commonly, in both. These parameters are relatively easy to measure in isolated muscle tissue, but in the intact heart direct measurements of these parameters are practically impossible to carry out. What is commonly done is to measure the *pressure* changes in the cavity of the heart and also the *changes in the volume* of blood in the heart (stroke volume). Both of these parameters should be measured for proper eval-uation of contractile function and even then, these measurements may not reflect accurately the contractile activity of the myocardium itself. Much uncertainty prevails about the definition of myocardial ''contrac-tility'' in the intact organism and the choice of criteria for assessing it.

Although one of the main functions of cardiac contraction is to maintain or adjust the stroke volume according to the needs of the tis-sues, stroke volume is a very poor index of the magnitude of the *con-tractile force* or the degree of *shortening* of heart muscle. For instance, if the ventricle is *acutely dilated,* the same stroke volume will be achieved by a lesser degree of shortening of the fibers (fig. 78) or the same degree of shortening will result in a greater stroke volume (fig. 79). If the left ventricle is assumed to be spherical, the volume of a sphere changes with the 3rd power of change in radius ($^4/_3 \pi \Delta r^3$) whereas the circumference changes (or shortens) with the first power of change in radius ($2 \pi \Delta r$). Also, if the left ventricle is dilated, the same stroke

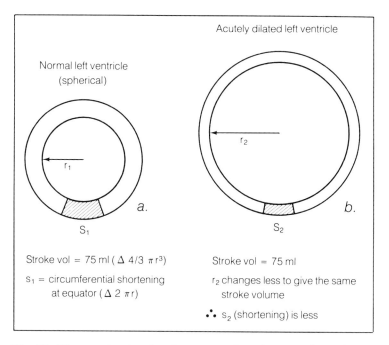

Fig. 78. Diagram showing that the same stroke volume can be achieved in a di-lated ventricle by a smaller degree of fiber shortening due to geometric considerations.

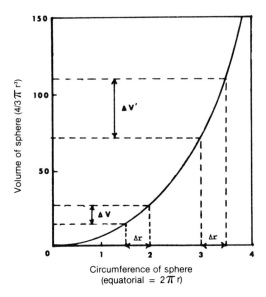

Fig. 79. Exponential relationship between the volume and circumference of a sphere at its equator. Assuming a spherical left ventricle, the same degree of circumferential shortening ($2\pi\Delta r$) produces a greater change in volume ($4/3\pi\Delta r^3$) in dilated, compared with nondilated ventricle. [Reproduced, with permission, from Badeer, H.S.: Cardiovascular adaptations in the trained athlete; in Lubich and Venerando: Sports Cardiology. (A. Gaggi, Publisher, Bologna 1980).]

volume and ventricular pressure requires a greater contractile force (tension) of the ventricular muscle in accordance with the equation of Laplace:

$$T = \frac{P \cdot r}{2}$$

(where T = tension in wall, P = transmural pressure, r = radius, when the ventricle is considered as a *single surface* spherical body). If there is a moderate increase in peripheral vascular resistance with constant heart rate and output (stroke volume), the contractile force or tension of the left ventricle will increase although the stroke volume is unchanged (ventricular pressure rises). All these situations indicate that *stroke volume is not a reliable index of ventricular contractility and should never be used for this purpose.*

Some of the criteria used to assess ventricular ''contractility'' are:
1.   Peak systolic ventricular pressure (fig. 80);

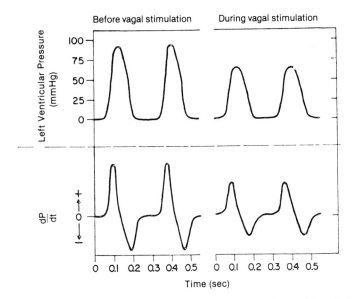

Fig. 80. A liquid-filled balloon is placed in a dog left ventricle, the heart paced at a constant frequency, and the coronaries perfused with blood. During vagal stimulation, heart rate and fiber length remain unchanged (isovolumic) keeping the two variables that can affect contractility constant. Note the decrease in left-ventricular peak pressure and dP/dt during vagal stimulation (negative inotropic effect on ventricular muscle). [Reproduced, with permission, from Berne, R.M.; Levy, M.N.: Cardiovascular Physiology; 4th ed. (C.V. Mosby Co., St. Louis, MO. 1981).]

2. Maximum velocity of blood in the aorta (fig. 81).
3. Ejection fraction (SV)/(EDV).
4. "Ventricular function curve" of Sarnoff [1955]. This curve plots external stroke work (= mean ejection pressure × stroke volume) of the left ventricle against left ventricular end-diastolic pressure when the latter is increased by increasing venous inflow into the ventricle (fig. 82). If the curve moves up, contractility is increased and vice versa.
5. Peak ventricular "power" during systole. (Power, as defined by physicists, is rate of doing work.) Maximal rate of doing external work during ejection:

$$= \max P \frac{dV}{dt} \quad \text{where P = pressure during ejection.}$$

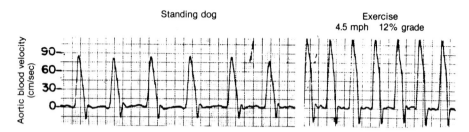

Fig. 81. During muscular exercise, cardiac sympathetic nerves are stimulated, increasing the velocity of ventricular contraction and ejection. The duration of ejection is shortened despite the small increase in stroke volume that occurs. As a consequence, peak aortic blood velocity increases. [Reproduced, with permission, from Rushmer, R.F.: Cardiovascular Dynamics; 3rd ed. (W.B. Saunders Co., Philadelphia 1970).]

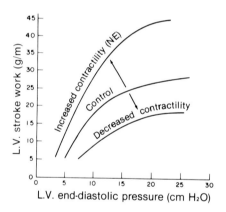

Fig. 82. Left ventricular function curve of Sarnoff. The administration of a positive inotropic agent (eg, norepinephrine) characteristically increases the stroke work at a *given* left ventricular end-diastolic pressure (or volume) and is said to increase contractility. If the ventricle performs less stroke work at a given end-diastolic pressure (or volume), contractility is said to be decreased (negative inotropic state). [Reproduced, with permission, from Braunwald, E.; Ross, J. Jr.; Sonnenblick, E.H.: Mechanism of Contraction of the Normal and Failing Heart (Little, Brown and Co., Boston, MA. 1967).]

6. Maximum dP/dt = maximum rate of change of left ventricular pressure (first derivative of ventricular pressure). This occurs toward the end of the isovolumic phase of systole (fig. 80). Several investigators have divided max dP/dt by other parameters of ventricular pressure to express contractility.

7. Maximum velocity of shortening of the contractile elements (V_{max}). This is the velocity of shortening of fibers extrapolated to zero load (theoretical maximum) (see Ch. 2, figs. 18–21).

According to some, V_{max} represents the "contractile" or "inotropic" state of the muscle. Early studies of Sonnenblick showed that changing the initial length of the muscle (Frank-Starling mechanism) does not change the V_{max}, although it affects the P_o (or isometric peak tension). In other words, increasing the initial length of the muscle will tend to increase the isometric peak force development but does not affect the "intrinsic" velocity of shortening at zero load (fig. 83). Hence, Sonnenblick [1962] claimed that initial muscle length or the Starling mechanism is not a determinant of myocardial "contractility." His arguments for this view are not very convincing (although most observers accepted it) and there is no reason to believe that maximum isometric force development (P_o) is not as good a criterion as V_{max}. It should be

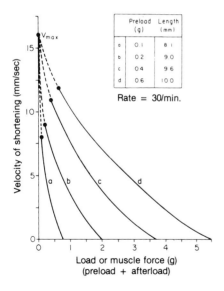

Preload (g)	Length (mm)
a	0.1	8.1
b	0.2	9.0
c	0.4	9.6
d	0.6	10.0

Rate = 30/min.

Fig. 83. Papillary muscle force-velocity measurements using an after-loaded isotonic experiment. Curves a–d correspond to increasing initial lengths below optimum initial length (L_o). When extrapolated toward the velocity axis, these curves merge into a common value of V_{max}. [Reproduced, with permission, from Pollack, G.H.: Maximum velocity as an index of contractility in cardiac muscle. Circ. Res. 26: 111 (1970). After Sonnenblick, E.H.: Force-velocity relations in mammalian heart muscle. Am. J. Physiol. 202:931 (1962).]

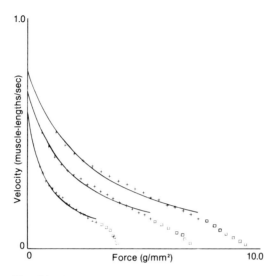

Fig. 84. Force-velocity curve from cat papillary muscle at three lengths on the ascending limb of the length-tension curve (lengths shorter than L_o). The three curves represent the best-fitting rectangular hyperbola calculated by the method of least sum of squared velocity values. Data points used in calculating the hyperbola curve are indicated by plus signs; the other points are open squares. [Reproduced, with permission, from Donald, T.C.; Unnoppetchara, K.; Peterson, D.; Hefner, L.L.: Effect of initial muscle length on V_{max} in isotonic contraction of cardiac muscle. Am. J. Physiol. *223:* 262 (1972).]

noted that more recent studies have not confirmed Sonnenblick's early work and that initial muscle length does in fact affect V_{max} (fig. 84) and also affects the rate of *tension* development (dT/dt) in isometric recordings (figs. 85 and 86).

In the intact heart, the Frank-Starling mechanism operates whenever the muscle fibers are *acutely* stretched, as in situations where there is an increased inflow into the ventricles *per cardiac cycle* (stroke-inflow). This may occur normally when changing from a standing to a recumbent position or when there is marked bradycardia with normal inflow/min, or artificially when large intravenous infusions of fluid or blood are given. Sarnoff [1958] labeled this type of intrinsic adjustment *heterometric autoregulation* (variable initial length). According to the current view, the functional significance of the Frank-Starling mechanism is to equalize the stroke volume or output of the two ventricles whenever a condition alters the venous inflow into one of the ventricles (eg, nonsteady states).

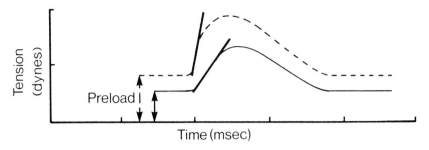

Fig. 85. Isometric recording. Note the effect of increasing initial muscle length (or tension) on rate of tension development (dT/dt) and on peak tension developed when total load and other factors are held constant. [Courtesy of Dr. Nora Laiken]

Sarnoff [1960] made further observations that were at variance with those of Starling and others. He noted that increased peripheral vascular resistance and aortic pressure, with constant heart rate and stroke volume in an isolated perfused heart, were *not* accompanied by an increase in left ventricular end-diastolic pressure and end-diastolic fiber length (fig. 87). In these experiments contractile force increased without an apparent increase in initial fiber length. He described this as *homeometric autoregulation* (also referred to as the ''Anrep'' effect). Its mechanism remains obscure. Homeometric autoregulation of the heart has not been demonstrated in studies of intact animals.

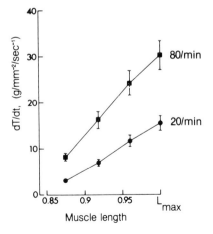

Fig. 86. Plot showing the increase in rate of tension development (dT/dt) with increasing initial muscle length at lengths below L_{max} (under steady state conditions for two frequencies of stimulation—20 and 80/min). [Reproduced, with permission of the American Heart Association, from Lakatta, E.G.; Jewell, B.R.: Length-dependent activation: its effect on the length-tension relation in cat ventricular muscle. Circ. Res. *40:* 251 (1977).]

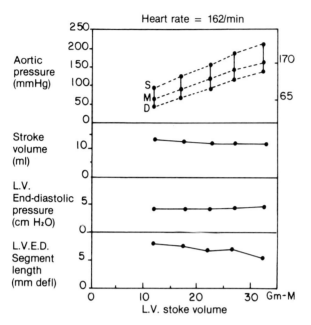

Fig. 87. Isolated supported heart. Constancy of left ventricular end-diastolic segment length as arterial resistance is increased (heart rate and venous inflow constant)—homeometric autoregulation. [Reproduced, with permission, from Sarnoff, S.J.; Mitchell, J.H.; Gilmore, J.P.; Remensnyder, J.P.: Homeometric autoregulation of the heart. Circ. Res. *8:* 1077 (1960).]

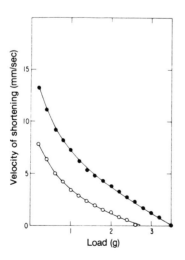

Fig. 88. The influence of norepinephrine (NE) on the force-velocity curve of isolated cat papillary muscle. Note the increase in both V_{max} and P_0. ○—○ Control; ●—● norepinephrine, 0.05 μg/ml. [Reproduced, with permission, from Braunwald, E.; Sonnenblick, E.H.; Ross, J. Jr.: Contraction of the normal heart; in Braunwald: Heart Disease—A Textbook of Cardiovascular Medicine. (W.B. Saunders, Philadelphia 1980).]

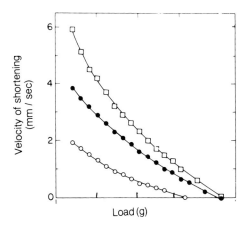

Fig. 89. Isolated cat papillary muscle driven at rates 20 (○—○), 30 (●—●), and 50 (□—□) times per minute. Note the increase in both V_{max} and P_0 as the frequency of stimulation is increased. [Reproduced, with permission, from Sonnenblick, E.H.: The mechanics of myocardial contraction; in Briller and Conn: The Myocardial Cell. (University of Pennsylvania Press, Philadelphia 1966.).]

An important determinant of "inotropic" state is the influence of sympathetic nerves or catecholamines. In an isolated papillary muscle preparation catecholamines increase both the V_{max} and the peak iso-metric tension (P_0) (fig. 88). In the in situ heart, sympathetic stimulation increases ventricular dP/dt and the ejection velocity and thereby reduces end-systolic and end-diastolic ventricular volumes provided total peripheral vascular resistance is not markedly increased (cardioacceleration per se may also play a role). A similar effect occurs on atrial muscle.

Another factor that may affect the "inotropic" state is the frequency of the heart beat. In the isolated papillary muscle preparation, increased frequency of stimulation increases V_{max}, but the effect on P_0 is variable, depending on the range of frequency (fig. 89). Kruta studied the influence of frequency of stimulation on *peak* force over the entire range of frequencies in *isolated strips* of heart muscle from different animals. The general pattern is shown in figure 90. This effect is referred to as the frequency-force relationship. The underlying mechanism is not understood although there are interesting hypotheses to explain it. Whether or not the *intact* heart shows a frequency-force relation is not certain. Kavaler found that the in situ, blood-perfused, dog papillary muscle did not show a frequency-force relation, but when the same

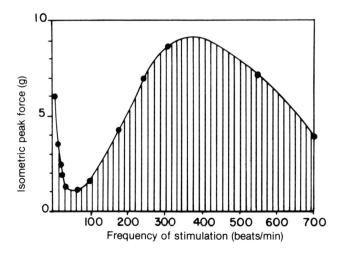

Fig. 90. Effect of frequency of stimulation on peak force of contraction (under steady states) of isolated atria of guinea pig at 35°C. Changes are believed to be related to restorative processes after each contraction that seems to have both a positive and a negative inotropic effect with different intensities and time course of decay. [Redrawn, with permission, from Kruta, V.: Sur l'activité rythmique du muscle cardiaque. I. Variations de la reponse mecanique en fonction du rythme. Arch. Int. Physiol. *45:* 332 (1937).]

muscle was isolated, it did show a frequency-force relation. He could not determine the factor responsible for the change!

An interesting observation is that premature beats very early in diastole have a potentiating effect on the force of the subsequent beat. This phenomenon is called *postextrasystolic potentiation*. It is not due solely to the greater filling and Frank-Starling mechanism as one might suspect, because it is seen in isolated strips of heart muscle and in the isovolumic left ventricle (balloon in left ventricle filled with liquid) where initial fiber length is held constant (fig. 91). Its mechanism remains obscure.

In addition to the above factors, contractility is affected by changes in the concentration of certain *cations* in the extracellular fluid. In the isolated papillary muscle, increasing Ca^{2+} from 2 to 5 mM/L increases both the V_{max} and the P_0 (fig. 92). Ca^{2+} plays an important role in E-C coupling. It is stored mainly in the sarcotubular system with some, in the mitochondria and is released into the sarcoplasm with the spread of the action potential. It is then pumped back into the sarcotubular system

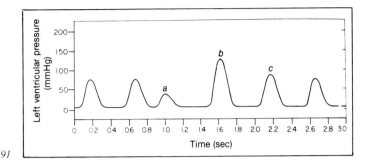

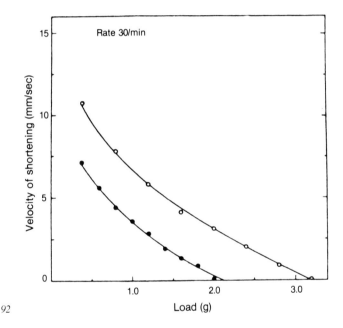

Fig. 91. Isovolumic dog left ventricle (fluid-filled balloon in left ventricle, LV). At (a) the ventricle is stimulated prematurely with a single shock. After a longer than normal pause, the beat b is stronger than normal (postextrasystolic potentiation). Note that the enhanced contractility persists for a short period (contraction c) before returning to the control value. [Reproduced, with permission, from Berne, R.M.; Levy, M.N.: Cardiovascular Physiology; 4th ed. (C.V. Mosby, St. Louis, MO. 1981).]

Fig. 92. The effect of an increase in $[Ca^{2+}]$ from 2.0 to 5.0 mM/liter on V_{max} and P_0 in the isolated papillary muscle. Temperature, 20°C. ●—● control; ○—○ Ca^{2+} 5mM/liter (rate 30/min). [Reproduced, with permission, from Sonnenblick, E.H.: Force-velocity relations in mammalian heart muscle. Am. J. Physiol. *202:* 931 (1962).]

and the mitochondria by an active process thus causing relaxation of the muscle fibers. The regulatory mechanisms are not known. Currently much work is in progress to understand the role of Ca^{2+} in regulating myocardial contractility. Toxic concentrations of Ca^{2+} in the extracellular fluid can stop the heart in a "contracted state" (calcium rigor). In the total absence of Ca^{2+}, heart muscle cannot contract, although action potentials can still be induced.

Potassium is another important cation. Increased extracellular potassium (normal = 4–5 mEq/L plasma) reduces the isometric peak tension of isolated papillary muscle. V_{max} is most probably reduced, but I have not been able to locate experimental data. Excess K^+ reduces not only the resting membrane potential (depolarizes) but also reduces the slope of the upstroke and amplitude of the cardiac action potential. Excess K^+ delays conduction from the atria to the ventricles and may cause com-

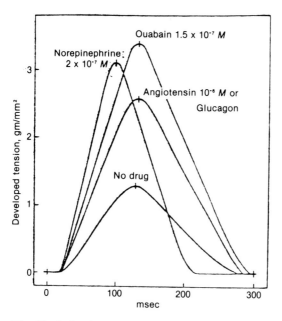

Fig. 93. Isolated cat papillary muscle contracting isometrically. Note the positive inotropic action of norepinephrine, ouabain, angiotensin, and glucagon. [Reproduced, with permission, from Koch-Weser, J.: Effects of beta-adrenergic stimulation and blockade on myocardial mechanics; in Kattus, Ross and Hall: Cardiovascular Beta Adrenergic Responses (University of California Press, Berkeley 1970).]

plete A-V block. Ultimately the heart stops in a somewhat relaxed state (complete depolarization). Clinically, plasma K^+ may increase sufficiently to cause cardiac disturbances (EKG and contractile changes) in cases of acute renal failure with uremia. A lesser degree of hyperkalemia may occur when adrenal cortical function is deficient (Addison's disease). In these conditions, K^+ excretion in the urine is diminished and accumulates in the blood.

Glucagon (from α-cells of the islets of Langerhans) in large doses increases the contractility of heart muscle. It increases cAMP but the receptor is not β-adrenergic. Also, angiotensin II increases peak contractile tension of isolated papillary muscle (fig. 93).

Severe arterial hypoxemia and acidosis tend to depress the contractility of the heart.

Temperature also affects myocardial contractility. Little work has been done on the effect of hyperthermia, although clinically, fevers are frequent occurrences. Many studies have been carried out on the effect of hypothermia on cardiac contractility. Cooling isolated heart muscle from 37°C to 20° or 25°C *increases* the peak isometric tension and the duration of contraction.

In disease, additional factors can alter myocardial contractility. Congestive heart failure is characterized by depressed contractility of ventricular muscle which is usually hypertrophied. Drugs like digitalis glycosides (ouabain, digoxin) are used to improve myocardial contractility (fig. 93).

In summary, it can be said that many factors can alter the "contractile" function of the normal myocardium and regulate stroke volume. Some mechanisms are intrinsic (heterometric or homeometric autoregulation) and others are extrinsic—through nerves, hormones, ions, or oxygen in the blood. It is difficult to say which of these are more frequently called into play than others in the intact normal heart.

Note. Experimentally, one factor that *does not* change the force of myocardial contraction is the intensity of the stimulus. The response of heart muscle does not vary with the intensity of the stimulus, if the stimulus is above threshold. This is called the "all or none law." The myocardium is a syncytium (acts as a single cell) from a physiologic standpoint, although the cells are histologically discrete and separate. Therefore, there is no spatial summation in the heart. This law should not be interpreted to mean that myocardial contractility is invariable.

References

Donald, T.C.; Unnoppetchara, K.; Peterson, D.; Hefner, L.L.: Effect of initial muscle length on Vmax in isotonic contraction of cardiac muscle. Am. J. Physiol. *223*:262–267 (1972).

Feigenbaum, H.: Echocardiography; 2nd ed. (Lea and Febiger, Philadelphia 1976).

Frank, O.: On the dynamics of cardiac muscle. [translated by Chapman, C.B., Wasserman, E. from Zeit. Biol. 32:370–447 (1895).] Am. Heart J. *58*:282–317; 467–478 (1959).

Guyton, A.C.; Jones, C.E.; Coleman, T.G.: Circulatory Physiology: Cardiac Output and Its Regulation; 2nd ed. (W.B. Saunders, Philadelphia 1973).

Jewell, B.R.: A reexamination of the influence of muscle length on myocardial performance. Circ. Res. *40*:221–230 (1977).

Katz, L.N.: The performance of the heart. Circulation *21*:483–498 (1960).

Koch-Weser, J.; Blinks, J.R.: The influence of the interval between beats on myocardial contractility. Pharmacol. Rev. *15*:601–652 (1963).

Lakatta, E.G.; Jewell, B.R.: Length-dependent activation: its effect on the length-tension relation in cat ventricular muscle. Circ. Res. *40*:251–257 (1977).

Patterson, S.W.; Piper, H.; Starling, E.H.: The regulation of the heart beat. J. Physiol. (London) *48*:465–513 (1914).

Rushmer, R.F.: Constancy of stroke volume in ventricular responses to exertion. Am. J. Physiol. *196*:745–750 (1959).

Rushmer, R.F.; Smith, O.; Franklin, D.: Mechanisms of cardiac control in exercise. Circ. Res. *7*:602–627 (1959).

Sarnoff, S.J.: Myocardial contractility as described by ventricular function curves: observations on Starling's law of the heart. Physiol. Rev. *35*:107–122 (1955).

Sarnoff, S.J.; Mitchell, J.H.; Gilmore, J.P.; Remensnyder, J.P.: Homeometric autoregulation in the heart. Circ. Res. *8*:1077–1091 (1960).

Sonnenblick, E.H.: Implications of muscle mechanics in the heart. Fed. Proc. *21*:975–990 (1962).

10   "External" Work of the Heart (Ventricles)

Heart muscle converts chemical energy into mechanical work by developing tension (force) and shortening of the muscle ("internal" mechanical work) and thereby performs work on the blood by imparting energy to it ("external" mechanical work). In liquids that are displaced from one position to another without undergoing onward flow, it can be shown that work done is equal to the *pressure exerted on the liquid times the volume of liquid displaced.* Remember that

Work = force × distance

In liquids, total force = pressure × area (L^2)
Substituting for force:
Work done on a liquid = pressure × area (L^2) × distance (L)
Since, area (L^2) × distance (L) = volume (L^3)
Work = pressure × volume (displaced)

If the liquid is given momentum, in addition to simple displacement, the kinetic energy of motion should be added to obtain the total work done on the liquid. The ventricles not only move the blood into the respective artery but also impart momentum to it. Thus the energy imparted to the blood takes the form of (a) pressure energy, and (b) kinetic (flow) energy. The units of work and energy are the same. In fact, energy may be defined as the capability of doing work.

External Work on Blood

1. Pressure-volume energy

$$\int_{V_1}^{V_2} P\,dV$$

P = *instantaneous* ventricular pressure *during ejection* over and above ventricular end-diastolic pressure.
V_1 = end-diastolic ventricular blood volume (EDV)
V_2 = end-systolic ventricular blood volume (ESV)
($V_1 - V_2$ = stroke volume)

2. Kinetic energy $= \frac{1}{2} M v^2$

$$\int_{V_1}^{V_2} \rho \frac{v^2}{2} \, dV$$

M = mass of blood ejected
v = *instantaneous* velocity at ventricular outflow tract during
    ejection
ρ = density of blood

Written more simply:

$$\text{External Stroke Work} = (\bar{P} \cdot \Delta V) + \left(\rho \frac{\overline{v^2}}{2} \cdot \Delta V \right) = \Delta V (\bar{P} + \rho \frac{\overline{v^2}}{2})$$

P = integrated mean of instantaneous ventricular pressure *during
    ejection* minus end-diastolic ventricular pressure.
ΔV = stroke volume
$\overline{v^2}$ = integrated mean of the *square of instantaneous velocity
    during ejection*

Accurate measurements of these variables (particularly velocity) are elaborate and calculations are usually much simplified, eg, instead of mean ventricular pressure *during ejection,* often mean aortic or pulmonary pressures are used.

Gravity has no influence on cardiac external work by acting *directly* on the blood and altering its energy content. The reason is that gravity has two opposing influences on energy content of blood: (a) weight of column of blood (gravitational pressure of column $= \rho g h$), and (b) gravitational potential energy, related to the distance from the center of gravity of the earth (energy content due to position $= \rho g h$). In the circulatory system, these two energies vary to the same extent in opposite directions so that their sum at any point in the system remains constant. On the other hand, gravity has a very significant *indirect influence* on cardiac external work by altering cardiac output and total peripheral resistance. Gravity affects the *transmural* pressure along the entire vascular tree as a result of the weight of the column of blood in the vessels. Since blood vessels are distensible, the caliber of arteries, arterioles, capillaries, venules, and veins are affected *passively* thereby altering total peripheral resistance, venous inflow and cardiac output—hence, cardiac work is altered as defined above.

Calculating the "External" Work of the Adult Human Heart Under Resting Conditions (Approximate). At this time the reader may find it useful to review the physical units presented in Appendix 1.

Stroke Work

Left Ventricle

1. Pressure-volume work
   $= 70 \text{ cm}^3 \cdot 11 \text{ cm Hg}$
   $= 70 \text{ cm}^3 \cdot 146,608 \text{ dynes/cm}^2 \qquad = 10,262,560 \text{ dyne} \cdot \text{cm (ergs)}$

2. Kinetic energy (ergs)
   $= 1/2 \cdot 70 \text{ g} \cdot (100 \text{ cm/sec})^2$ *Note:* 100 cm/sec is mean *ejection* velocity at aortic orifice
   $= 35 \text{ g} \cdot 10,000 \text{ cm}^2/\text{sec}^2$ calculated from stroke volume, cross-sectional area of orifice,
   and ejection time (70 cm³, 3 cm², and 0.22 sec respectively)
   $= 350,000 \text{ g} \cdot \text{cm}^2/\text{sec}^2 \text{ (ergs)} \qquad = \quad 350,000 \text{ ergs}$

   Total LV work/stroke $\qquad = 10,612,560 \text{ ergs} = 1.06 \text{ joules}$

Right Ventricle

Ejection pressure $= 1.5 \text{ cm Hg} = 1.5 \cdot 13.6 \cdot 980 = 19,992 \text{ dynes/cm}^2$

1. $70 \text{ cm}^3 \cdot 19,992 \text{ dynes/cm}^2 \qquad = 1,399,440 \text{ dyne} \cdot \text{cm (ergs)}$
2. Same as that of left ventricle $\qquad \underline{350,000 \text{ ergs}}$

   Total RV work/stroke $\qquad = 1,749,440 \text{ ergs} = 0.175 \text{ joule}$

   Work of *two* ventricles/beat $\qquad = 12,362,000 \text{ ergs} = 1,2362 \text{ joules or newton-meters}$
   $(1 \text{ joule} = 10^7 \text{ ergs})$

Work of Two Ventricles Per Minute

$= \text{Stroke work} \cdot \text{heart rate/min}$
$= 1.2362 \text{ joules} \cdot 75 \text{ beats/min} = 92.7 \text{ joules or newton-meters/min}$
Per day??
Per lifetime ???

Comments

(a) Kinetic energy factor *normally* is a small fraction of total work and is usually omitted in calculations. However, if cardiac output is high it constitutes a significant portion of total work and its omission will lead to serious error. (b) The work of the left ventricle is about seven times greater than that of the right ventricle because of the difference in mean pressure during ejection between the aorta and the pulmonary artery.

11 Work and Energy Expenditure of the Heart

The traditional method of calculating cardiac work is to determine the work done *on the blood*. This is referred to as the "external" work and is described in the preceding chapter.

A useful way to represent the external work is to plot the stroke-work on the basis of changes in ventricular pressure and volume with each heart beat (fig. 94). This graph neglects the kinetic factor, which is usually a small fraction. The pressure-volume curves (diastolic and systolic) may shift under different physiologic and pathologic states and the stroke-work area may also vary with circumstances.

Very often, it is desirable to know the energy expenditure of a machine to do a given external work and thus estimate the "mechanical efficiency" of the machine. This is particularly difficult in living machines doing mechanical work because living cells consume energy even when not doing external mechanical work (energy is used to maintain cell life). The immediate source of energy is from the breakdown of ATP, which is rapidly restored by the phosphate derived from creatine phosphate (CP) and by the oxidation of carbohydrates and fats (to a small extent, the amino acids) through the Krebs cycle (oxidative phosphorylation) (fig. 95). Fatty acids are an important substrate for the heart under normal conditions, and in the fasting state they constitute the most important fuel for the heart (65%).

Respiratory quotient (R.Q. = $\dot{V}CO_2$ produced/$\dot{V}O_2$ consumed) is usually slightly below 0.8.

Energy expenditure of the heart, under steady states, is almost entirely oxidative because in the total absence of oxygen, heart muscle rapidly fails to contract within a minute or so (anaerobic glycolysis provides a very limited supply of ATP). Thus, under steady states heart muscle must restore its energy during each cardiac cycle by taking up oxygen, carbohydrates and lipids from the coronary capillary blood. These facts justify the measurement of O_2 consumption of the heart as an index of overall cardiac energy expenditure.

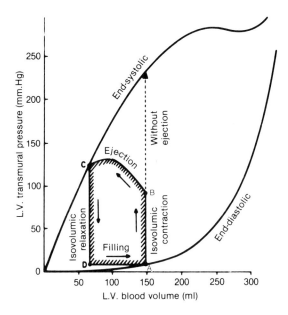

Fig. 94. Pressure-volume changes in the left ventricle of a normal resting adult with an aortic pressure of 130/90 mmHg, stroke volume of 80 ml and an end-systolic (residual) volume of 70 ml.

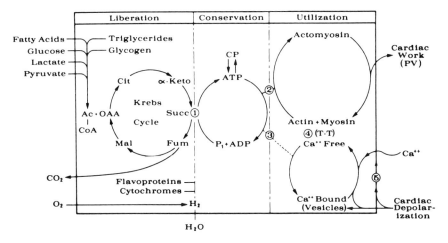

Fig. 95. Diagram showing the energy conversions in cardiac muscle. [Reproduced, with permission, from Olson, R.E.; Dhalla, N.S.; Sun, C.N.: Changes in energy stores in the hypoxic heart. Cardiology 56: 114 (1971–72).]

The oxygen uptake of any organ can be determined by the Fick principle. $\dot{V}O_2 = \dot{Q}\ (aO_2 - \bar{v}O_2)$. $\bar{v}O_2$ is the oxygen content of the *mixed* venous blood from all veins draining the organ. This is impossible to obtain if the organ has many veins, but in the heart of large animals and of man, the coronary sinus drains *chiefly the left ventricular muscle,* and sinus blood can be obtained with a catheter. The normal oxygen uptake of the heart is about 8–10 ml O_2/100 g left ventricular muscle per minute. Of this amount, about 2 ml O_2/100 g per minute is for maintaining the viability of the *noncontracting heart muscle.*

The classic teaching about energy expenditure of the heart is that it depends on the "external" work of the heart. This *erroneous* view was derived from controlled experiments in the Heart-Lung preparation where *one* variable was altered, keeping the others constant (Evans 1918). For instance, with constant heart rate and blood pressure, increasing output up to a limit increases oxygen consumption of the heart linearly but the

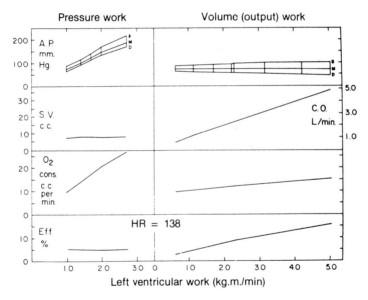

Fig. 96. Experiment to show the different effects of aortic pressure (peripheral resistance) and cardiac output on myocardial oxygen consumption (heart rate constant). Note that increasing aortic pressure (pressure work) increases O_2 consumption of the heart much more than would increasing cardiac output (volume work). AP = aortic pressure; SV = stroke colume; CO = cardiac output; Eff % = mechanical efficiency; HR = heart rate. [Reproduced, with permission, from Folkow, B.; Neil, E.: Circulation. (Oxford University Press, Oxford 1971).]

slope is low grade (much less than 45°). Thus, "volume" work per se causes only a slight increase in oxygen usage (fig. 96). On the other hand, with constant heart rate and output (constant stroke volume), increasing peripheral resistance (hence arterial pressure) up to a limit increases the oxygen uptake of the heart also linearly but the slope is about 45° (tangent of angle = 1) (fig. 96). Hence, "pressure" work of the heart consumes much more energy than "volume" work (other things being equal).

Careful analysis of these observations indicates that if the two variables, stroke volume and blood pressure, change at random in different directions and to different degrees (heart rate being constant), the oxygen consumption of the heart may or may not vary directly with the external work. The reason is that pressure and stroke volume have equal importance in determining "external" cardiac work ($W = \bar{P} \cdot \Delta V$), but with regard to oxygen consumption arterial pressure is *more* influential or potent than stroke volume. Hence, oxygen uptake of the heart may not necessarily vary directly with "external" work. Sarnoff [1958] proved this experimentally by increasing cardiac output and simultaneously decreasing the peripheral resistance to a *lesser extent* (heart rate being constant) and showed that increased external work of the heart is accompanied by a decrease in myocardial oxygen consumption (fig. 97).

The second reason why oxygen consumption of the heart *per minute* may not correlate with external work per minute is that heart rate has an influence on myocardial oxygen consumption. Doing the same pressure-volume work per minute, a given heart consumes *less oxygen per minute at a slow rate* than at a fast rate.

In the third place, oxygen uptake of the heart and *external* work do not correlate well if there are changes in ventricular end-diastolic volume (eg, compliance). An *acutely* dilated heart, beating at the same rate and stroke volume against the same blood pressure as a nondilated heart, will consume more oxygen. This is related to the greater contractile tension that must be developed in the dilated heart because of the law of Laplace. Laplace's formula indicates that contractile tension in the wall of a thin-walled cylinder or sphere is dependent not only on transmural pressure but also on the *radius of curvature*.

For cylinder $T = P \cdot r$

For sphere $T = \dfrac{P \cdot r}{2}$ } very thin-walled.

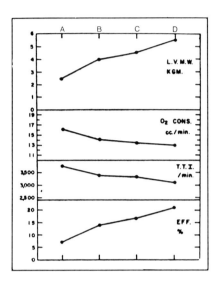

Fig. 97. Lack of correlation between external work (= CO × aortic pressure) and myocardial oxygen consumption when both cardiac output and peripheral resistance change in different directions and degrees (heart rate constant). CO is increased and at the same time peripheral resistance is decreased to a lesser extent so that cardiac work is increased. Note the difference in cardiac oxygen consumption, which correlates with TTI/min but not with external cardiac work/min. LVMW = left ventricular minute work. [Reproduced with permission from Sarnoff, S.J.; Braunwald, E.; Welch, G.H. Jr.; Case, R.B.; Stainsby, W.N.; Macruz, R.: Hemodynamic determinants of oxygen consumption of the heart with special reference to the tension-time index. Am. J. Physiol. *192:* 148 (1958).]

The larger the radius, the greater must be the wall tension for a given pressure (fig. 98). A practical application of this relationship is seen in the construction of containers subject to high internal pressures, such as gas tanks. To avoid bursting, these are built with small radii of curvature. In contrast, containers under low pressure (eg, milk truck tanks) are built with much larger radii (fig. 99).

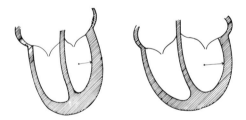

Fig. 98. Increasing the radius (r) of a given heart (eg, acute dilation), keeping pressure (P) and wall thickness (δ) constant, increases the contractile tension or stress (T). Laplace formula for a thin-walled sphere is $\overline{T} = \overline{P} \times \overline{r}/2\delta$ where T = mean stress in the wall during systole; $\overline{P}$ = mean transmural pressure during systole; $\overline{r}$ = mean radius of curvature; $\overline{\delta}$ = mean thickness of wall.

Milk truck

Truck carrying
gas under pressure

Fig. 99. The application of Laplace's equation in industry. A milk truck can use a single tank with a large radius because the pressure is low, whereas a truck carrying gas under high pressure must use tanks that have a small radius of curvature. [Reproduced, with permission, from Katz, A.M.: Physiology of the Heart. Raven Press, New York 1977).]

Sarnoff [1958] found that myocardial oxygen consumption correlates better with the *area under the aortic pressure curve during the period of ejection* than with the external work of the heart. He called this area the *tension-time index* or TTI. Note that he measured pressure and not myocardial tension, but called it tension-time. It may be better described as pressure-time product (fig. 100). This concept explains why "pressure" work is more oxygen consuming than "volume" work.

More recent studies have reported that even better correlation occurs if heart rate times peak myocardial *tension* is calculated by applying the Laplace formula to the left ventricle (fig. 101).

Further work has shown that in addition to the development of tension, the extent of myocardial *shortening* and the *velocity* of shortening (V_{max}) are determinants of myocardial oxygen consumption. These variables are significantly affected by the cardiac sympathetic nerves and circulating catecholamines. These agents increase V_{max}, dP/dt, and P_o (isometric peak tension). Some authors claim that sympathetics and catecholamines increase cardiac energy expenditure also by *direct metabolic* action independent of the contractile effects. This so-called calorigenic action is believed to be small.

A scheme to indicate the components of myocardial oxygen consumption is shown in figure 102.

In Conclusion. The traditional view that cardiac oxygen consumption depends on "external" work must be dropped because it holds only

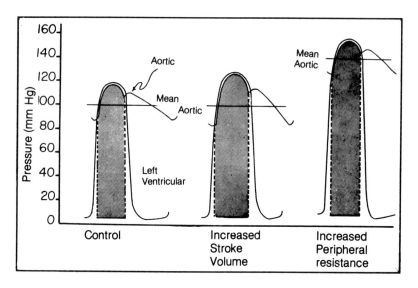

Fig. 100. Diagram showing the effect of increasing cardiac output and peripheral resistance on the TTI/beat when heart rate is constant. The greater area under the curve during ejection (TTI) when peripheral resistance is increased is believed to account for the greater myocardial oxygen consumption. The role of the Laplace law (changes in end-diastolic volume) on tension development is not evident in this type of analysis. Shaded area = TTI/beat. [Reproduced, with permission, from Badeer, H.S.: Work and energy expenditure of the heart. Acta Cardiol. *24:* 227 (1969).]

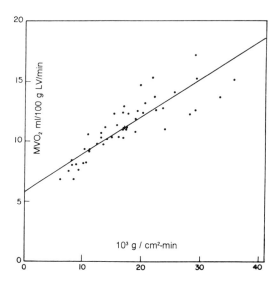

Fig. 101. Graph showing the good correlation between myocardial oxygen consumption per minute ($M\ddot{V}_{O_2}$) and the product of peak ventricular *tension* (calculated using Laplace, $T = P \cdot r$) and heart rate/min. [Reproduced, with permission, from McDonald, R.H., Jr.; Taylor, R.R.; Cingolani, H.E.: Measurement of myocardial developed tension and its relation to oxygen consumption. Am. J. Physiol. *211:* 667 (1966).]

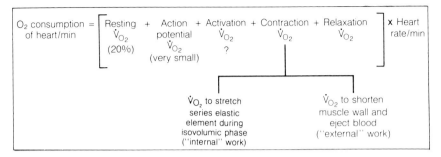

Fig. 102. O_2 consumption of heart.

under special circumstances. When many variables change at random, oxygen consumption may not correlate with external work.

At present it is believed that the important determinants of myocardial oxygen consumption are:

1. Magnitude of myocardial *tension* development (during the isovolumic period to stretch the series elastic elements and during ejection; affected by aortic pressure and the radius of curvature of the ventricle). ⎫
2. Velocity of contraction (dP/dt, V_{max}). Influenced by sympathetic nerves, catecholamines, heart rate, blood calcium, temperature, and so on. ⎬ Major
3. Frequency of heart beats/min (acute effect). ⎭
4. Degree of fiber shortening (stroke volume) plays a relatively minor role.
5. Temperature of heart muscle has a direct metabolic effect in addition to changing the heart rate and velocity of contraction.

For clinical purposes at the bedside, Katz [1958] has suggested the use of the product of *mean arterial pressure/cycle* × heart rate/min, called the *HR·BP index*. This is a crude estimate because it neglects end-diastolic size, stroke volume, velocity of contraction, and so on. This index plus X-ray studies of heart size should give a *fairly good idea* about the metabolic strain on the heart in disease states in clinical practice.

The use of aortic pressure (by Sarnoff and Katz) instead of the ventricular pressure is justified in practically all conditions because aortic pressure and ventricular pressure during ejection are almost equal. How-

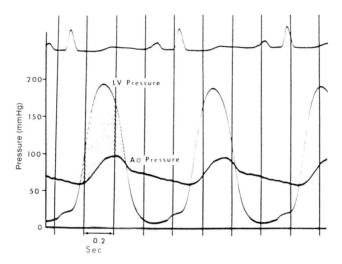

Fig. 103. Record of left ventricular pressure and aortic pressure in a subject with aortic valvular stenosis. The shaded area shows the marked pressure gradient between the left ventricle and the aorta (compare this with the normal gradient shown in figs. 42 and 43). [Reproduced, with permission, from Dodge, H.T.; Kennedy, J.W.: Cardiac output, cardiac performance, hypertrophy, dilatation, valvular disease, ischemic heart disease and pericardial disease; in Sodeman and Sodeman: Pathologic Physiology: Mechanisms of Disease; 5th ed. (W.B. Saunders Co., Philadelphia 1974).]

ever, in aortic stenosis in which there is a large ventriculo-aortic gradient of pressure (fig. 103), aortic pressure cannot be used to estimate ventricular energy expenditure.

It is believed that myocardial energy expenditure has an important bearing on the development of cardiac hypertrophy and perhaps ultimate failure!

References

Badeer, H.S.: Work and energy expenditure of the heart. Acta Cardiol. (Brussels) *24:*227–241 (1969).

Braunwald, E.: Control of myocardial oxygen consumption. Physiologic and clinical considerations. Am. J. Cardiol. *27:*416–432 (1971).

Burton, A.C.: The importance of the size and shape of the heart. Am. Heart J. *54:*801–810 (1957).

Evans, C.L.: The velocity factor in cardiac work. J. Physiol. (London) *52:*6–14 (1918–19).

Katz, L.N.: The performance of the heart. Circulation *21:*483–498 (1960).

Katz, L.N.; Feinberg, H.: The relation of cardiac effort to myocardial oxygen consumption and coronary flow. Circ. Res. *6:*656–669 (1958).

McDonald, R.H., Jr.; Taylor, R.R.; Cingolani, H.E.: The measurement of myocardial developed tension and its relation to oxygen consumption. Am. J. Physiol. *211:*667–673 (1966).

Sarnoff, S.J.; Braunwald, E.; Welch, G.H., Jr.; Case, R.B.; Stainsby, W.N.; Macruz, R.: Hemodynamic determinants of oxygen consumption of the heart with special reference to the tension-time index. Am. J. Physiol. *192:*148–156 (1958).

12 Cardiac Reserve

The heart has the ability to deal with increasing loads without failing, up to a limit. By load is meant the amount of blood (or fluid) in the ventricles before systole begins (*preload*) and the arterial pressure against which the ventricles pump blood (*afterload*). *Acute* increases in cardiac load occur frequently in everyday life, eg, muscular activity, intake of food and drinks, exposure to heat and cold, emotional upsets, and so forth.

The heart deals with such *acute* increases in load by:
1. Cardioacceleration - stretch S-A node, reflexes, hormones, psychic influences, temperature changes, and so on.
2. Increased "contractility" to maintain or to augment stroke volume: (a) intrinsic mechanisms such as homeo- and heterometric autoregulation, Ca^{2+}, inotropic effect of heart frequency, and so on; and (b) extrinsic mechanisms through nerves and hormones (sympathetics, catecholamines changing V_{max} and dP/dt, or thyroid hormone, and so on.)

The maximal work the heart is capable of performing without failing is called *work "capacity" of the heart*.

Work capacity − work done at rest = work reserve.

Unfortunately, there is no way of quantifying the work capacity of the *normal intact heart*. Muscular exercise is the best way of maximizing cardiac load and it is commonly believed that the severest exercise does not exhaust the work capacity of the heart. Some cardiac reserve always remains. Fatigue of skeletal muscles or central synapses puts an end to the exercise.

In *chronic* overloading of the heart, another reserve mechanism comes into play. The heart muscle *hypertrophies*. Fibers become thicker by adding more myofibrils and also become longer by adding more sarcomeres. Cardiac myocytes do not multiply in the adult. These changes serve to increase the total force across the thicker wall of the cardiac chamber.

The stimulus to hypertrophy seems to be a chronic increase in *mean contractile tension/beat* or, better, a chronic increase in *energy expenditure/beat*. Hence, dilated hearts with larger radii of curvature, irrespective of cause, tend to hypertrophy as a result of increased contractile tension $(T = P \cdot r)$. Likewise, increased pressure in a cardiac chamber causes hypertrophy (eg, stenosis of valves, hypertension, and so forth). On the other hand, increased stroke volume increases tension-time *per beat* only slightly and, therefore hypertrophy is not very prominent, specially if pressure increases do not occur. When hypertrophy is fully established, the energy expenditure is restored toward normal and the muscle ceases to hypertrophy (so-called stable stage of hypertrophy). Overall coronary blood flow per unit mass of hypertrophied myocardium is usually within normal limits under resting conditions. Most studies on "contractility" of hypertrophied myocardium find it is *depressed*.

Athletic training also causes a mild or moderate hypertrophy of the heart ("physiologic" hypertrophy) not to exceed about 500 g. It is probably related to repeated increases in *stroke* energy expenditure and/or the resting *bradycardia* that develops, both of which tend to augment stroke energy expenditure. "Contractility" of such hearts seems to be better than that of the hearts of nontrained animals. There is evidence that discontinuation of athletic activity may partly restore the heart muscle mass, particularly when hypertrophy is slight.

References

Badeer, H.S.: Myocardial blood flow and oxygen uptake in clinical and experimental cardiomegaly. Am. Heart J. *82:*105–119 (1971).

Badeer, H.S.: Development of cardiomegaly. Cardiology (Basel) *57:*247–261 (1972).

Cohen, J.; Shah, P.M.: Cardiac hypertrophy and cardiomyopathy. Circ. Res. *34/35:* (suppl. II): (1974).

Meerson, F.Z.: The myocardium in hyperfunction, hypertrophy and heart failure. Circ. Res. *24/25* (suppl. II): (1969).

The Blood Vessels

13 Hemodynamic Principles

Hemodynamics is the study of the relationships among pressure, resistance and flow in the cardiovascular system. One of the commonest errors in classic teaching is the statement that ''fluid flows from higher pressure to lower pressure.'' This implies that the driving force causing flow is the difference in pressure between two points. This statement is only partly true because it can explain observations of flow under certain special circumstances but fails to explain flow under all circumstances. Also, failure to distinguish between pressure gradients caused by gravity on a column of liquid (so-called hydrostatic pressure, better called gravitational pressure) and pressure gradients necessary to cause flow in a system of tubes (flow or dynamic pressure) has caused much confusion and erroneous thinking. An example of the difficulty in explaining flow on the basis of pressure gradient *alone* is the following: a beaker filled with fluid has a higher pressure at the bottom than at the top, yet there is no flow toward the top. The reason is that flow actually depends not on pressure difference but on *total energy difference*. Total energy of a liquid at any point is defined by *Bernoulli's equation*, which is the main equation in hydraulics.

$$\underset{\text{(per unit volume)}}{\text{Total energy (E)}} = \underset{\substack{\text{dynamic P} + \text{gravitational P} \\ (\propto \dot{Q} \cdot R) \quad\quad (\pm\rho gh)}}{\overbrace{\text{Pressure energy}}} + \underset{(\pm\rho gh)}{\substack{\text{Gravitational} \\ \text{potential} \\ \text{energy}}} + \underset{(\tfrac{1}{2}\rho v^2)}{\substack{\text{Kinetic} \\ \text{energy}}} + \substack{\text{Thermal} \\ \text{energy}}$$

where $\dot{Q}$ = flow per unit time; R = downstream resistance to flow (hydraulic resistance); ρ = density of liquid; g = acceleration of gravity; h = height of liquid column from a given horizontal reference plane; v = velocity of flow.

All items on the right side of the equation are expressed per unit volume and are *interconvertible* except thermal energy which cannot be converted back into the others. Applying Bernoulli's equation to the

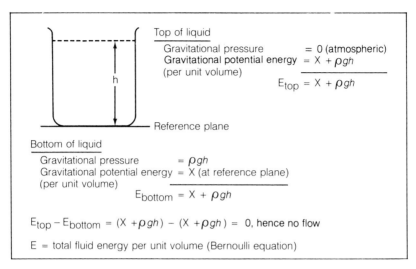

Fig. 104. Application of Bernoulli's theorem to explain why in a beaker filled with a liquid, there is no flow from the higher pressure at the bottom to the lower pressure at the top.

example of the beaker it becomes clear why there is no flow from the bottom to the top if the temperature of the liquid is uniform (fig. 103).

Another example where pressure gradients cannot explain flow whereas energy gradients, based on Bernoulli's equation, can explain is illustrated in figure 105. In the constricted zone, velocity is higher than elsewhere because the cross-sectional area is smaller, hence kinetic energy per unit volume ($\frac{1}{2}\rho v^2$) is greater. As a result, dynamic pressure at the narrow zone is lower than at the distal wider zone, although the sum of pressure energy and kinetic energy is greater in the constricted zone compared to the sum of pressure and kinetic energy distal to it. The sum of these two energies continually declines along the tube as indicated by the slope of the line. This energy is converted into heat due to the friction between the particles of the liquid (viscosity). Note that in this case gravitational pressure energy and gravitational potential energy of the liquid do not play a role because the tube is placed horizontally. This model also illustrates the fact that dynamic pressure energy can be converted into kinetic energy and vice versa in a given system. A similar situation may occur clinically in coarctation of the aorta where the low pressure tends to collapse the arterial wall at the narrow segment of the artery.

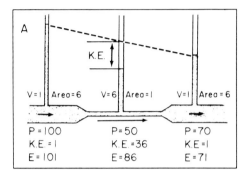

Fig. 105. Diagram to illustrate the validity of Bernoulli's equation. Flow occurs along a gradient of *total* fluid energy (E) but not along a gradient of pressure. In the narrow region of the tube the kinetic energy of the liquid ($\frac{1}{2}$ mv²) is increased because of the increase in velocity (v = $\dot{Q}$/area). This occurs at the expense of pressure energy which is reduced markedly. In the model, pressure at the constricted region is taken to be 50 units, whereas in the distal wider region, the pressure is taken to be 70 units (some kinetic energy is reconverted into pressure). Thus flow from the constricted to the distal wider region cannot be explained on the basis of pressure gradient but can be explained on the basis of total energy gradient (86 units versus 71 units). [Reproduced, with permission, from Burton, A.C.: Physiology and Biophysics of the Circulation; 2nd ed. (Year Book Medical Publishers, Chicago 1972).]

The dynamics of flow in the vascular system is exceedingly complex because blood has corpuscles, the vessels are distensible, flow is pulsatile; vessels branch, unite and curve; the circuit is closed; gravity acts on blood; and so forth. To simplify, we will use a model with homogeneous liquid flowing in a straight rigid circular tube of uniform diameter placed horizontally with steady flow (open circuit). If the linear velocity of flow is not high, the liquid layers slide over each other smoothly, giving rise to *laminar or streamline flow*. The molecules immediately next to the wall do not move at all and those in the center move fastest. The velocity profile is parabolic (fig. 106). The hydrodynamic pressure against the wall can be measured by side tubes at various points along the tube. The height of the column of liquid at each point represents the *transmural* pressure at that point, P_1, P_2, P_3, etc. (fig. 107). Transmural pressure is the pressure difference between the inside and outside at any given point (net pressure across wall). In this model the outside pressure is atmospheric, whereas in the body the pressure outside the vessels may not be atmospheric. The difference in dynamic pressure between two points along the tube or vessel (P_1–P_2 or P_1–P_3)

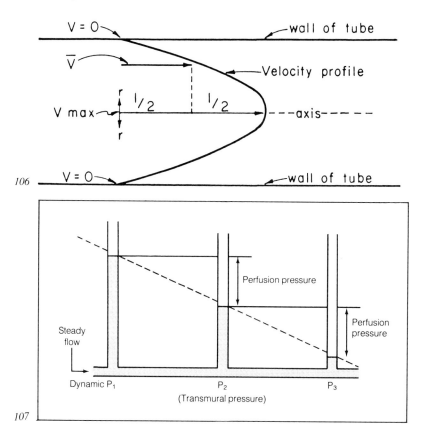

Fig. 106. Diagram of the velocity profile (parabolic) of a viscous liquid (Newtonian) flowing in a steady streamline manner in a rigid tube. The average velocity (v) is one-half the maximal velocity (v_{max}) at the center of the tube. Note that the velocity immediately immediately adjacent to the wall is zero. [Reproduced, with permission, from Little, R.C.: Physiology of the Heart and Circulation; 2nd ed. (Year Book Medical Publishers, Chicago 1981); modified from McDonald, D.A.: Blood Flow in Arteries; 2nd ed. (Edward Arnold Publishers, London 1974).]

Fig. 107. Diagram showing the difference between transmural and perfusion pressure.

is called the *perfusion pressure* or *driving pressure*. It represents the loss of energy in moving the liquid from one point to the other and is indicated by the pressure drop between the points. The terms give the somewhat false idea that the flow is due entirely to the pressure gradients rather than energy gradients. However, the pressure concept is truly applicable in a single horizontal tube of uniform bore because the other factors of the Bernoulli equation remain the same along the tube. Note that transmural pressure at a given point and perfusion pressure between two points along a tube are entirely different concepts in hydraulic systems. If the outlet of the tube is tilted down, gravity acts on the column of liquid, causing an increase in transmural pressure along the rigid tube but the perfusion pressures between P_1, P_2, and P_3 remain unchanged, provided the flow rate is kept constant.

The amount of fluid flowing per unit time is called the volume flow or flow rate ($\dot{Q}$). The relationship between perfusion pressure and flow rate was studied by a French physician named Poiseuille. Using rigid capillary tubes of uniform diameter and a homogenous liquid with streamline flow, he found that P_1-P_2 varies:

1. Directly with the distance between P_1 and P_2 (L).

Therefore,

$$P_1-P_2 \propto L \text{ (constant } \dot{Q}, \text{ r and } \eta)$$

where r = internal radius; η = viscosity of liquid.

2. Directly with the flow rate, $\dot{Q}$

$$P_1-P_2 \propto \dot{Q} \text{ (constant L, r, } \eta)$$

3. Directly with the viscosity of liquid, η

$$P_1-P_2 \propto \eta \text{ (constant L, } \dot{Q},r)$$

4. Inversely with the 4th power of radius:

$$P_1-P_2 \propto \frac{1}{r^4} \text{ (constant L, } \dot{Q}, \eta)$$

Therefore,

$$P_1-P_2 \propto \frac{L\eta\dot{Q}}{r^4} = k\frac{L\eta\dot{Q}}{r^4},$$

k was found to be $\dfrac{8}{\pi}$.

$$\text{So, } P_1-P_2 = \frac{8}{\pi} \cdot \frac{L\eta\dot{Q}}{r^4} \quad \text{or} \quad \boxed{\dot{Q} = \frac{(P_1-P_2)\pi r^4}{8 L \eta}}$$

This relationship is known as *Poiseuille's law or equation*. It can also be derived mathematically. Note that viscosity (η) has a time factor and its unit is the *poise,* which is equal to 1 dyne sec/cm^2. It should be emphasized that P_1-P_2 in Poiseuille's equation *excludes the difference in gravitational pressure between the two points.*

It is customary to refer to $8L\eta/\pi r^4$ as the *resistance* (R) between points P_1 and P_2. Hence, Poiseuille's equation may be simplified to $P_1-P_2 = R \dot{Q}$ or

$$\dot{R} = \frac{P_1 - P_2}{\dot{Q}}.$$

(This equation is analogous to Ohm's law; $I = E/R$ or $E = IR$; volts = amperes $\times$ ohms.)

Note that resistance in such a *single tube* depends on three variables:

R $\propto$ directly with length or distance between two points
R $\propto$ directly with the viscosity of liquid
R $\propto$ inversely with r^4

In the vascular system the situation is different because, among other things, the tubes repeatedly branch. The branches constitute tubes that are placed "in parallel" and the total resistance (R_T) of such a system is:

$$\frac{1}{R_T} = \frac{1}{R_1} + \frac{1}{R_2} + \frac{1}{R_3} \text{ , and so on.}$$

If each branch has the same radius and length as the single tube, the more the branches, the less will be the *total R* (fig. 108). Likewise, it follows that in a system of parallel tubes the total resistance must be less than that of any individual component. Thus, in any vascular bed total R varies with four factors:

$$R_T \propto \frac{L \text{ (vessels in series) } \eta}{\bar{r}^4 \cdot \text{No. of parallel vessels}},$$

In other words $R_T \propto$ directly with length in series; $R_T \propto$ directly with viscosity; $R_T \propto$ inversely with 4th power of mean internal radius; $R_T \propto$ inversely with number of open parallel vessels. The vascular factors (length, radius and number of parallel vessels) are sometimes lumped together as the "geometric" factors in contrast to the blood factor of "viscosity." Because Poiseuille's equation was developed for a *single tube* we have modified it for circulation as follows:

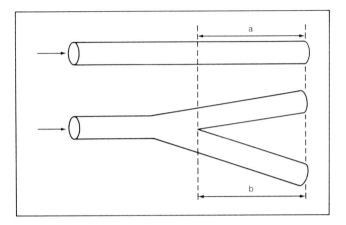

Fig. 108. Two branching vessels (b) having the same diameter and length as that of a single vessel (a) will have a total resistance one-half that of the single vessel.

$$\dot{Q} = \frac{(P_1 - P_2)\, \pi r^4}{8\, L\, \eta} \cdot \text{No. of parallel vessels}$$

If resistances are placed "in series," the total R is equal to the sum of the individual resistances. Some authors prefer to use the reciprocal of resistance, which is designated as *conductance*. Conductance = 1/R.

At this point it is important to distinguish between *volume flow* and *velocity* of flow. Flow ($\dot{Q}$) is volume/unit time, whereas velocity is distance/unit time. By definition:

$$\text{Mean velocity } (\bar{v}) = \frac{\dot{Q}}{\text{cross-sectional area}} \text{ or in a circular tube } \bar{v} = \frac{\dot{Q}}{\pi r^2}$$

If we substitute for $\dot{Q}$ from the Poiseuille's equation we obtain:

$$\bar{v} = \frac{(P_1 - P_2)\, \pi r^4}{8\, L\, \eta} \cdot \frac{1}{\pi r^2} = \frac{(P_1 - P_2)\, r^2}{8\, L\, \eta}$$

Consequently, if (P_1-P_2), L and η are *kept constant*, increasing the radius *increases* the velocity of flow. This may appear paradoxical because increasing r increases the cross-sectional area and this might be expected to decrease the velocity. That would be true if flow ($\dot{Q}$) were to remain unchanged, but we see from the Poiseuille equation that when r increases and (P_1-P_2) is kept constant, flow $(\dot{Q})$ would increase very markedly (by r^4). Because flow increases by r^4 and the cross-sectional area increases by r^2, velocity, which is flow/cross-sectional area, increases by r^2 (r^4/r^2).

This situation occurs in the body when blood vessels dilate *locally* without causing any change in arterial blood pressure. Both blood flow and velocity increase during local vasodilation in a tissue and vice versa.

Applicability of Poiseuille's Equation to the Circulatory System

Strictly speaking, Poiseuille's equation cannot be applied with any degree of accuracy to the cardiovascular system for the following reasons:
1. Blood has corpuscles and its viscosity is not a constant value in different parts of the system.
2. Blood flow in arteries is pulsatile (inertia effects) and that requires more energy than steady flow.
3. Blood vessels are distensible and, therefore, change their diameter and resistance with changes in transmural pressure.
4. Veins are not circular and that affects their resistance to flow.
5. At the root of the aorta and the pulmonary artery the flow is turbulent. Turbulent flow consumes more energy.
6. The importance of kinetic energy as a factor affecting flow is entirely overlooked (see Bernoulli's equation).
7. Diameter of vessels continually changes as one proceeds distally.
8. Curving of vessels causes additional loss of energy.
9. Branching and anastomosing create unknown complications, and so forth.

However, despite these serious limitations, it is still possible to use the Poiseuille equation in a general way by taking into account the role of each variable in the equation, eg, diameter is much more potent (r^4) than aortic pressure (P_1) in altering blood flow to tissues. This is the way blood flow is generally regulated and is highly effective.

References

Badeer, H.S.; Rietz, R.R.: Vascular hemodynamics: deep-rooted misconceptions and misnomers. Cardiology (Basel) *64*: 197–207 (1979).

Burton, A.C.: Physiology and Biophysics of the Circulation; 2nd ed.; pp. 97–100; 104–106 (Year Book Medical Publishers, Chicago 1972).

Caro, C.G.; Pedley, T.J.; Schroter, R.C.; Seed, W.A.: The Mechanics of the Circulation (Oxford University Press, New York 1978).

14 Mean Pressure and Mean Blood Velocity in the Vascular System

Mean pressure is highest in the aorta and progressively falls along the systemic vessels. By the time blood reaches the right atrium most of the energy of the blood is dissipated as frictional heat and the pressure remaining in the atrium is only a few mm Hg. The drop in pressure is due to the resistance of the vascular tree to the flow of blood. The mean pressure curve is characteristically nonlinear, depending upon the *total* resistance of the different parts of the vascular system (fig. 109). The greatest drop of pressure occurs in the arterioles; next in the capillaries. Pressure drops very little in the large arteries and veins.

The greater drop of pressure in arterioles as compared with that of the capillaries is due to a complex relationship of the variables that determine total resistance of vessels:

$$R \propto \frac{L \text{ (individual vessels) } \eta}{r^4 \cdot \text{No. of parallel vessels}}$$

Arterioles have greater individual length than capillaries and the "effective" viscosity of blood in arterioles is greater; also, they are fewer in number (fig. 110). All these tend to increase total resistance. On the other hand, they have larger radii which tend to decrease resistance. Apparently, the three factors that tend to increase resistance *outweigh* the effect of larger radii and the net influence is greater *total* resistance of arterioles than that of capillaries.

The mean velocity of blood in the vascular system presents a different curve. It is highest in the aorta, lowest in the capillaries, and intermediate in the veins and vena cava (fig. 111). The reason is the following: total blood flow through the different regions of the systemic circuit is the same under steady states and is equal to the cardiac output ($\dot{Q}$). Thus:

Total $\dot{Q}a_a$ = total $\dot{Q}_c$ = total $\dot{Q}_v$ a = arteries; c = capillaries; v = veins.

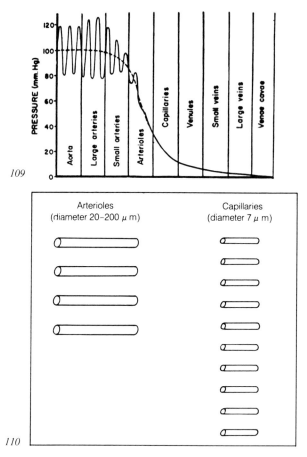

Fig. 109. Changes in mean dynamic pressure (excluding gravitational) in different parts of the systemic vascular tree. [Reproduced, with permission, from Guyton, A.C.: Function of the Human Body; 4th ed. (W.B. Saunders Co., Philadelphia 1974).]

Fig. 110. The role of diameter, length and number in determining the total resistance of arterioles and capillaries.

Since velocity (v) = Total $\dot{Q}$/Total x-area and total $\dot{Q}$ is the same in all three regions, velocity of blood at these regions of the circuit varies *inversely* with the *total* cross-sectional area at those regions. When arteries branch, the sum of the cross-sectional area of the branches is greater than that of the parent artery. Likewise, when venules unite to form larger veins, the total cross-sectional area of the venules is greater

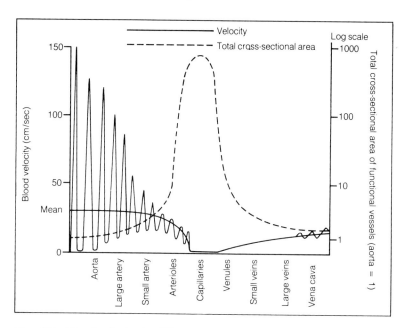

Fig. 111. Blood velocity in different parts of the systemic vascular circuit (total blood flow being the same).

than that of the larger veins. Consequently, the capillaries have the largest total cross-sectional area and the velocity of blood is lowest. The sum of cross-sectional areas of the two venae cavae is about twice that of the aorta, hence the velocity of blood in each vena cava is about one-half that of the aorta. A similar situation exists when a river passes through a narrow gorge where velocity becomes high; when the river reaches a wide bed the velocity may be hardly noticeable.

The reader should not confuse this situation with the effect of *localized* vasodilation on the velocity of blood flow in those vessels. As was discussed earlier, when vessels dilate, $\dot{Q}$ does not remain constant, but increases markedly (by r^4), whereas cross-sectional area increases only moderately (by r^2); since,

$$v = \frac{\dot{Q} \quad \propto r^4}{\text{x-area} \propto r^2} \qquad v \propto r^2$$

References

Badeer, H.S.; Rietz, R.R.: Vascular hemodynamics: deep-rooted misconceptions and misnomers. Cardiology (Basel) *64*:197–207 (1979).

Burton, A.C.: Physiology and Biophysics of the Circulation; 2nd ed.; pp. 97–100, 104–106 (Year Book Medical Publishers, Chicago 1972).

Caro, C.G.; Pedley, T.J.; Schroter, R.C.; Seed, W.A.: The Mechanics of the Circulation (Oxford University Press, New York 1978).

15 Pressure and Flow in Arteries

Direct measurements of pressure in the cardiovascular system are carried out by introducing a catheter or needle into any part of the system and connecting these to a pressure recording device (transducer) filled with an anticoagulant solution. One side of the transducer is exposed to *atmospheric pressure* commonly designated as ''zero'' pressure. Thus, all cardiovascular pressure measurements register the pressure of the blood inside the heart or vessels versus the outside atmospheric pressure. Since the force of gravity acts on the blood contained in the vascular system, pressure in blood vessels has two components: (a) dynamic or flow pressure and (b) gravitational pressure. Because gravity plays no role in causing blood flow in the *closed* vascular system, measurements of vascular pressure should exclude gravitational pressure. It has been noted that the right atrial pressure near the tricuspid valve (likewise the left atrial) is *almost* constant (close to zero or atmospheric) under different postural states (standing, lying down or even head-down position). This region is referred to as the *hydrostatic indifferent point* (HIP). Hence, the level of the tricuspid is taken as the *zero reference point for all circulatory* (dynamic) *pressure measurements*. The reason for such a finding is complex and seems to be related to the elastic properties of the vessels above and below the heart as demonstrated in physical models (Gauer). In practice, the gravitational component of vascular pressure is excluded by placing the *transducer at the level of the tricuspid*. This is so because the catheter or tube connected to the transducer is fluid filled and gravity acts equally on this fluid thereby neutralizing the gravitational pressure of the column of blood in the vessel (similar to that which occurs in a siphon). In the following discussion pressure always refers to the *dynamic* pressure created by the contractile activity of the ventricles (unless otherwise stated). This pressure is often erroneously described as ''hydrostatic'' pressure. There is nothing static about it.

The pressure in the aorta and large arteries rises and falls with each ejection of blood. The peak pressure occurs during systole and is called *systolic* (or maximum) *pressure*. The lowest pressure occurs at the end of diastole or rather at the *end* of isovolumic contraction and is called *diastolic* (or minimum) *pressure*. The difference between systolic and diastolic pressure is known as the *pulse pressure*. The area under the pressure curve during a cycle divided by the duration of the cycle gives the *true mean pressure* (area may be determined by graphic integration or by a planimeter; integration can also be achieved by electronic means). In the peripheral arteries, the true mean pressure is approximately equal to diastolic + ⅓ pulse pressure. Systolic + diastolic divided by 2 gives the *arithmetic* mean pressure, which in the aorta happens to be the same as the true mean pressure (fig. 112).

In clinical practice arterial pressure is determined by indirect methods. The most frequently used method is the auscultatory, which is described in Appendix 2. With this method the average value of dynamic pressure in the brachial artery of young persons at rest is about 120/80 mmHg. However, there is a wide variation of values in different individuals with a normal distribution curve. Any systolic pressure above

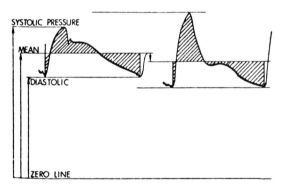

Fig. 112. True mean arterial pressure in the aorta (*left*) as compared with that of the iliac artery (*right*). In the aorta true mean pressure happens to be equal to the arithmetic mean of systolic and diastolic pressures (the shaded area above the midline is equal to the sum of the two shaded areas below the midline). In the peripheral vessel, despite the increase in pulse pressure, there is a fall in mean pressure approximately equal to the diastolic plus one-third of the pulse pressure. [Reproduced, with permission, from Gauer, O.: Kreislauf des Blutes; in Landois-Rosemann: Lehrbuch der Physiologie des Menschen; 28th ed. (Urban and Schwarzenberg, Munich 1960).]

150 mmHg or diastolic pressure above 100 mmHg (at any age) may be classified as systemic hypertension. Clinical experience has shown that high diastolic pressure is more detrimental to blood vessels than high systolic, leading to thickening of media and intima of arterioles and to atherosclerosis of large arteries. These changes tend to cause rupture or thrombosis (clotting of blood in vivo) of the affected vessels or cause interference with blood flow (ischemia).

The *mean* aortic pressure varies with the factors that appear in the simplified Poiseuille's equation.

$$P_1 - P_2 = \dot{Q} \cdot R$$

P_1 = mean aortic pressure
P_2 = mean right atrial pressure
$\dot{Q}$ = cardiac output (flow rate/min)
R = resistance of *entire* systemic circuit

P_2 is close to zero under most physiologic circumstances, hence the equation is often further simplified to $P = \dot{Q} \, R$ (not an accurate equation). Under these circumstances, aortic *transmural* pressure (P_1) and systemic *perfusion* pressure ($P_1 - P_2$) are numerically *about* equal, and factors determining perfusion pressure are used to assess variables that determine arterial (transmural) pressure. From this it is evident that *mean systemic arterial pressure* depends on two basic factors: (a) output of the heart ($\dot{Q}$) or (flow rate), and (b) total peripheral resistance (TPR) of systemic vessels.

Cardiac Output. Increasing output increases mean arterial pressure if TPR is constant and vice versa. In a *rigid* system of tubes with steady streamline flow the relationship between flow and perfusion pressure is linear and passes through the origin as predicted by Poiseuille's formula (fig. 113). However, in the vascular system the relationship is complex because the vessels are distensible, the smooth muscle reacts actively to changes in transmural pressure, effective viscosity of blood changes with flow, in certain organs tissue pressure may change with blood flow, vasodilator metabolites from tissues may be washed out, and so forth. All these variables tend to alter vascular resistance in vivo when (flow) pressure is altered. P and R are *interdependent variables*. Below a certain perfusion pressure there is no flow through vessels. This is called "critical closing pressure" by Burton. Here the equilibrium of forces in the wall of small vessels is disturbed·and the vessels close completely (fig. 113). In most organs (skeletal muscles, heart, kidney, brain, intes-

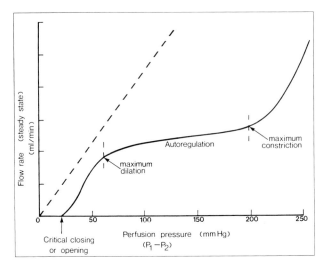

Fig. 113. Effect of perfusion pressure on flow rate in a rigid tube perfused with a Newtonian fluid (---) and in a vascular bed (muscle) perfused with blood (——). Note the autoregulation of steady state blood flow within certain limits of perfusion pressure in the circulation through muscle.

tines, etc.) above a certain perfusion pressure the steady state flow re-mains practically constant and independent of perfusion pressure. This is called *autoregulation of blood flow* (fig. 113). There are two major hypotheses to explain autoregulation of blood flow. One is the so-called *myogenic* hypothesis which states that increased perfusion pressure in-creases the transmural pressure and stretches the vascular smooth mus-cle. Stretch of smooth muscle causes it to contract. This narrows the vessels and increases vascular resistance which roughly balances the in-crease in $P_1 - P_2$, whereby flow remains constant $[(\dot{Q} = (P_1 - P_2)\uparrow/R\uparrow)]$. The second is the *metabolic* hypothesis. According to this concept, the metabolic products of tissues continuously diffuse out of cells and have a *local vasodilator* action thereby adjusting blood flow to the metabolic rate of tissues. In the above situation, the tissue has a constant metabolic rate and there is a steady production of vasodilators which diffuse into the blood and are carried away at a certain rate. When perfusion pres-sure is increased, there is an immediate increase in blood flow. This washes away the metabolites at a faster rate thereby reducing their con-centration in and around the blood vessels of the tissue. As a result, the vessels constrict increasing their resistance and restoring the blood flow. This mechanism operates until the maximal contractile response of vascular smooth muscle is reached, after which autoregulation breaks down (eg, above 200 mm Hg in the figure).

These two hypotheses to explain autoregulation will be further considered in Chapter 19 under the subject of peripheral vascular control.

Systemic TPR. This is offered chiefly by the arterioles, but the capillaries also contribute to a certain extent. Resistance is indicated by the *pressure drop* along vessels.

Vascular resistance cannot be measured directly; it can be *calculated* if $P_1 - P_2$ and $\dot{Q}$ are known. $R = P_1 - P_2 / \dot{Q}$. The units of pressure and flow are arbitrary. If pressure is expressed in dynes per square centimeter and flow in cubic centimeters per second, then R is in dynes/cm^2/cm^3/sec or dyne sec cm^{-5}. This is known as the absolute unit of resistance (CGS system). It has no virtue over other units. An other unit is the PRU (peripheral resistance unit). (Some authors express $\dot{Q}$ in ml/min).

$$1 \text{ PRU} = \frac{1 \text{ mmHg}}{1 \text{ ml/sec}}.$$

With this method, one can calculate the R of any portion of the vascular bed if the P_1 and P_2 at any two points and the *total* flow between those points are known.

Since organs vary markedly in their total mass and, therefore, in the number of parallel vessels, it appears more meaningful to compare the resistance of the vessels in different organs by considering the *blood flow per unit mass of tissue* (ml/sec · 100 g tissue) rather than the total organ flow. This may be referred to as the "specific" vascular resistance of an organ or tissue.

$$\text{Specific } R = \frac{P_1 - P_2}{\dot{Q} \text{ per unit mass of tissue}}$$

This would provide a crude idea about the functional vascularity of a tissue in relation to mass. The lower the specific R, the greater would be the vascularity. An example of a low specific vascular resistance is that of the kidneys as compared with an organ of similar weight, such as the heart. Both organs weigh about 300 g, but blood flow through the two kidneys is about five times greater, although perfusion pressure is about the same in both.

Problem. Calculate the systemic TPR and the pulmonary TPR in man in absolute units and in PRUs.

Note. The above treatment of pressure and flow is essentially for a steady (nonpulsatile) flow. Although this does not prevail in the circulatory system, the hemodynamic principles discussed above are adequate for practical purposes. A more sophisticated treatment of pressure and flow in the arterial system involves the concepts derived from alternating electrical currents. This involves complex problems of *vascular impedance* (instead of resistance) which has inertial, compacitive and resistive components. Such treatments are extremely difficult and beyond our scope.

Viscosity of Blood

The resistance to flow in the vascular system is mostly due to the viscosity of blood (η). Viscosity is described as the internal friction in moving fluids and is expressed as the ratio of shearing stress (displacing force) to the rate of shear (or strain).

$$\eta = \frac{\text{shearing stress (force)}}{\text{shearing strain (displacement)}}$$

When η is constant under different pressure and flow relations at a given temperature, the fluid is said to be Newtonian. When η changes with pressure and flow, the fluid is described as non-Newtonian. Blood is a non-Newtonian liquid. The cgs unit of viscosity is the *poise* (in honor of Poiseuille).

1/100 poise = 1 centipoise
Water at 20°C has a η of 1 centipoise (at 37°C = 0.7 centipoise)
Plasma at 37°C has a η of 1.2 centipoise
Blood at 37°C has a η of 2–10 centipoises depending upon hematocrit, size of tube, flow rate, etc. (average = 5 centipoises or 0.05 poise). This is about seven times that of water at 37°C.

Blood viscosity shows some peculiarities and varies with several factors.

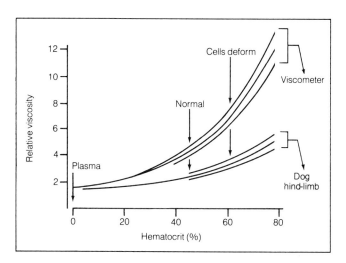

Fig. 114. Diagram showing the effect of hematocrit on the relative viscosity of blood (compared with that of water) in a viscometer with a tube radius greater than 1 mm, compared with the effect in blood vessels of dog's hindlimb. [Reproduced, with permission, from Burton, A.C.: Physiology and Biophysics of the Circulation; 2nd ed. (Year Book Publishers, Chicago 1972).]

Hematocrit. The hematocrit is the ratio of the volume of red blood cells to the volume of a given sample of blood. Normally it is about 45%. Increased hematocrit increases η, but the relationship is nonlinear both in vitro and in vivo (fig. 114). In anemia (eg, hematocrit of 30%) the effective viscosity is reduced somewhat but in polycythemia (eg, high altitude residents with hematocrit of 60%) the effective viscosity is increased to a greater extent. In some congenital heart defects the hematocrit may go up to 75%.

Diameter of Vessel and Velocity of Flow. In glass tubes with a *diameter* greater than 2 mm (radius 1 mm), the viscosity of blood is independent of tube diameter provided the flow rate is sufficiently high. At lower tube diameters, η becomes less as the diameter is reduced (called the Fähraeus-Lindqvist effect) (fig. 115). The cause is said to be the accumulation of RBC in the axial stream (axial accumulation or streaming). This process is believed to occur in the arterioles and would serve to reduce friction and the resistance to flow in these vessels (tends

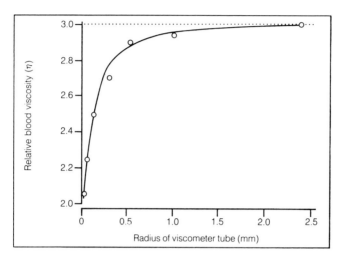

Fig. 115. The effect of viscometer tube radius below 1 mm on the relative viscosity of blood (η). Note the decrease in viscosity in tubes below 1 mm radius. For tubes with a radius larger than 1 mm, relative viscosity is not affected by tube radius. These effects are believed to be related to the axial accumulation of red blood cells in tubes with a radius below 1 mm. [Reproduced, with permission, from Burton, A.C.: Physiology and Biophysics of the Circulation; 2nd ed. (Year Book Medical Publishers, Chicago 1972).]

to lower arterial pressure and the work of the heart). Blood going to arterioles that come off at a large angle (eg, afferent arteriole of glomerulus) tends to have a lower hematocrit than blood in the main vessel (*plasma skimming*). In large vessels the cell free zone is about 5 μm and the cell poor zone is about 5–20 μm. Therefore, plasma skimming is of no importance in large arteries.

In *very narrow glass tubes,* the diameter of blood capillaries (eg, diameter 6 μm), the viscosity of blood approaches that of plasma. Here, the passage of red cells (which are very readily deformed) in single file with plasma between individual red cells seems to reduce the friction even more. In tubes with a diameter of 6 μm, the viscosity of blood is almost independent of hematocrit and is close to that of plasma (fig. 116).

The in vitro viscosity of blood tends to decrease as the velocity of flow increases. As the velocity increases, there seems to be greater axial accumulation of red cells, which reduces the overall friction between the layers (physical basis is complex and not well understood).

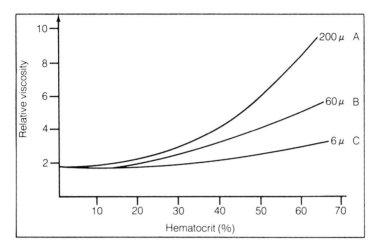

Fig. 116. The effect of hematocrit on relative viscosity of blood flowing at high velocity through rigid tubes of very small *diameter*. A = 200 μm; B = 60 μm; and C = 6 μm. A and B simulate the size of arterioles and C has about the diameter of capillaries. Note that with a normal hematocrit of 40%, blood in the 6-μm tube diameter (eg, capillary) has a lower relative viscosity than in tubes with diameters of 60 μm or 200 μm (eg, arterioles). [Reproduced, with permission, from Folkow, B.; Neil, E.: Circulation (Oxford University Press, Oxford 1971).]

Temperature. Cooling the blood increases its viscosity and vice versa. In hypothermia of 23°C, η is increased by about 40%. This tends to reduce blood flow if arterial pressure is unchanged.

Elasticity and Distensibility of Central Arteries

When a solid or semisolid is subjected to an external force it changes its shape which is regained when the force is removed. This is described as the elastic property. For linear or filamentous structures (eg, wire, rope, etc.) the relation between the force/unit cross-sectional area of the material (stress) and the elongation relative to original length (strain) is called the *modulus of elasticity (Young's modulus)*.

$$\begin{array}{c}\text{Elastic modulus} \\ \text{(dynes/cm}^2)\end{array} = \dfrac{\text{tensile stress (force/unit area)}}{\dfrac{\text{tensile strain } (\Delta L)}{(L)}}$$

If the elastic modulus is constant with varying degrees of stress, the material is said to obey Hooke's law. This is true for most metals, rubber and other materials (fig. 117). However, most living tissues do not obey Hooke's law (non-linear relationship). Note that *steel has a much higher elastic* modulus than rubber (for a 10% elongation, 2×10^{11} vs 4×10^6 dynes/cm^2). This is so because of the way physicists have defined and calculated elastic modulus (strain is in the denominator). The popular concept of elasticity is opposite to that of physicists— rubber is more ''elastic'' (popular concept = strain/stress).

In hollow viscera, like blood vessels or the heart, the measurement of stress and strain is difficult, if not impossible. Hence, one measures the transmural pressure across the wall in relation to changes in volume of contents. This gives the ''pressure-volume'' curve. If one plots the change in volume (Δv) in relation to the change in pressure (Δp), the relationship is described as *compliance* (same as in lungs).

$$\text{Compliance} = \frac{\Delta v}{\Delta p}$$

Some call this the absolute volume distensibility which has the disadvantage of giving values that vary with the original size of the organ.

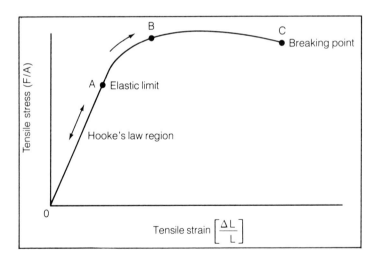

Fig. 117. The stress-strain diagram obtained from a wire with increased tension produced by an external force. [Reproduced, with permission, from Sears, F.W.; Zemansky, M.W.: University Physics; 2nd ed. (Addison-Wesley Publishing Co., Boston, MA. 1955).]

To avoid this, one may use the percent increase in volume in relation to increase in pressure. This has been called percent volume distensibility (Burton, 1972) or simply *distensibility* (Guyton, 1981). (Analogous to specific compliance of lungs).

$$\text{Distensibility } (\%) = \frac{\dfrac{\Delta v}{V} \times 100}{\Delta p} = \frac{\Delta v}{V \cdot \Delta p} \times 100$$

The pressure-volume curves of cardiovascular structures are non-linear and vary with age and pathologic factors. For the isolated aorta of man see figure 118.

The importance of the pressure-volume curve of central arteries is that it affects the systolic-diastolic pressure levels and the oscillations of flow in arteries and small vessels with each cardiac ejection. The volume of blood ejected into the aorta per beat is accommodated in two ways: (a) by driving some blood through the arterioles into the capillaries, called the *systolic run-off* (estimated to be about 20% of stroke volume) and (b) by distending the central arteries which store about 80% of the stroke volume. During diastole this volume of blood drives an equal volume through the arterioles (under steady states), described as the *diastolic run-off* (= stored volume).

Distensibility of central arteries serves to damp (reduce) the changes in pressure and flow in arteries by permitting a lower systolic pressure and a higher diastolic pressure than otherwise. The large central arteries act as a "compression chamber" or "windkessel." (In electrical circuits this is achieved by a "capacitor" which filters out the oscillations in voltage, provided there is enough resistance ahead.)

The pressure oscillations in peripheral arteries as compared with the aorta show characteristic differences. In distal arteries such as the femoral, brachial, radial, dorsalis pedis, and so on, the systolic is higher and the diastolic slightly lower than in the aorta. The upstroke is sharper and the dicrotic wave is at a lower level in the descending portion of the curve (fig. 119). The *true* mean pressure is slightly lower in the periphery, as might be expected. The explanation for these changes is controversial but the following factors are believed to operate: (a) pressure waves are reflected backward from the bifurcations of arteries in the periphery; (b) high frequencies of vibrations in the aorta travel at a greater velocity than the lower frequencies, causing changes along the arterial tree; and (c) higher frequencies are more easily damped as they

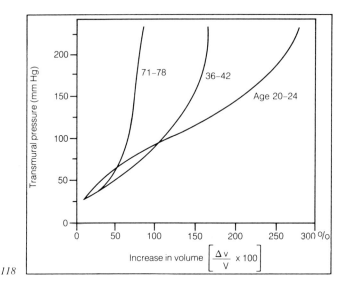

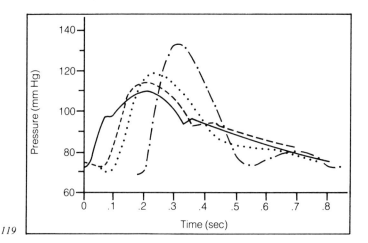

Fig. 118. Pressure-volume curves of human aortas obtained at autopsy from different age groups. [Redrawn from Hallock, P.; Benson, I.C.: Studies on the elastic properties of human isolated aorta. J. Clin. Invest. *16:* 595 (1937).]

Fig. 119. Pressure oscillations in different parts of the systemic arteries in man. Note the increase in pulse pressure as one proceeds from the aorta to the periphery (for explanation see text). — 0 cm withdrawal; --- 20 cm withdrawal; ··· 40 cm (femoral); ·—·—· dorsalis pedis. [Reproduced, with permission, from Guyton, A.C.: Textbook of Medical Physiology; 5th ed. (W.B. Saunders Co., Philadelphia 1976).]

travel. All these and others add up to cause the changes described above. Some explain it on the basis of greater rigidity of peripheral arteries.

References

Gauer, O.H.; Thron, H.L.: Postural changes in the circulation; in Hamilton and Dow: Handbook of Physiology; sect. 2; Circulation, vol. III; pp. 2409–2439 (American Physiological Society, Washington, D.C. 1965).

Guyton, A.C.: Textbook of Medical Physiology; 6th ed. (W.B. Saunders, Philadelphia 1981).

Hallock, P.; Benson, I.C.: Studies on the elastic properties of human isolated aorta. J. Clin. Invest. *16:*595–602 (1937).

Johnson, P.C.: Autoregulation of blood flow. Circ. Res. *14/15,* suppl. I: (1964).

McDonald, D.A.: Blood Flow in Arteries; 2nd ed. (Williams and Wilkins, Baltimore 1974).

Whittaker, S.R.F.; Winton, F.R.: The apparent viscosity of blood flowing in the isolated hind limb of the dog and its variation with corpuscular concentration. J. Physiol. (London) *78:*339–369 (1933).

16 The Arterial Pulse

Changes in transmural pressure in the aorta with each cardiac ejection cause changes in vascular diameter that travel to the periphery as the pulse wave. In the intact organism the changes in the diameter of peripheral arteries are slight if no external pressure is applied on the vessel (*about 5% increase*). If pressure is applied on an artery, as in taking the pulse, the movements of the wall are augmented. Movements of the aortic wall travel with a velocity that varies with the rigidity of the arteries. In the aorta the velocity is about 4–5 m/sec; in medium-sized arteries it is 5–14 m/sec (more rigid); and in small arteries, 20 m/sec or more.

The velocity of the pulse wave that travels *in the arterial wall* must not be confused with the velocity of the blood within the arteries, which moves much slower. In the aorta of man the mean velocity *of blood* during the *entire cardiac cycle* is about 30 cm/sec (can be calculated from the stroke volume, cross-sectional area of aorta, and the duration of *cardiac cycle*) under resting conditions. Note that in calculating the kinetic work of the heart, the mean aortic velocity *during the period of ejection* must be used. This velocity is much higher—about 100 cm/sec—because the ejection period is much shorter than the entire cardiac cycle (fig. 120).

Aortic blood flow or velocity has been recorded with electromagnetic or ultrasonic flowmeters. The results show that there is practically no forward flow in the ascending aorta during diastole (fig. 120). This is partly due to the rigid cuff of the flowmeter placed around the aorta, thereby preventing the distension and storage of blood during systole in this segment. As one proceeds along the arterial system, the pulsations in flow velocity decrease continuously, and practically disappear in the systemic capillaries (recently capillary flow is said to be normally pulsatile). The conversion of pulsatile flow into a uniform flow in capillaries is due to the cooperation of three factors: (a) complete closure of the

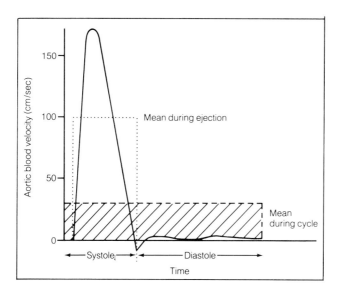

Fig. 120. Diagram showing the difference between mean aortic orifice blood velocity during ejection and mean aortic velocity during the entire cardiac cycle.

aortic valve (provided there are no communications between the aorta and the pulmonary artery or between a large artery and a large vein); (b) proper elasticity or distensibility of central arteries in relation to stroke volume; and (c) proper resistance of small arteries and arterioles.

These principles are illustrated in a model shown in figure 121 where a hand bulb with two, one-way valves, is squeezed at a certain rate and

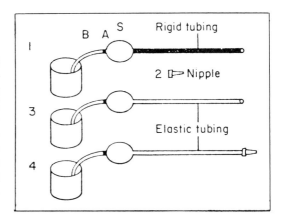

Fig. 121. Model showing the factors involved in converting pulsatile flow in arteries to steady flow in capillaries. Note that in this model the role of the closure of the outflow valve is overlooked. [Reproduced, with permission, from Brobeck, J.R.: Best and Taylor's Physiological Basis of Medical Practice; 9th ed. (Williams & Wilkins, Baltimore 1973).]

draws water from a basin. If the outflow tube is rigid, with or without a nipple, the fluid comes out intermittently. If the tube is elastic but there is no resistance ahead, again the fluid runs out in spurts. On the other hand, elastic tubing with a nipple (resistance) can convert the intermittent output of the pump into a steady flow out of the tube. *All three factors* are necessary and must be in proper proportions in order to achieve uniform velocity of flow in the capillaries. The physiologic value of such flow is to ensure a *steady* supply of nutrients and removal of waste products from tissue cells.

Clinically, systemic capillary pulsation may occur in: (a) aortic regurgitation; (b) large patent ductus arteriosus or A-V fistula; and (c) peripheral arteriolar dilation (eg, severe anemia, hyperthyroidism, exposure to heat, and so forth).

Excessive aortic rigidity is a possibility but is not listed as a cause in clinical medicine.

References

Gauer, O.H.; Thron, H.L.: Postural changes in the circulation; in Hamilton and Dow: Handbook of Physiology; sect. 2; Circulation, vol. III; pp. 2409–2439 (American Physiological Society, Washington, D.C. 1965).

Guyton, A.C.: Textbook of Medical Physiology; 6th ed. (W.B. Saunders, Philadelphia 1981).

Hallock, P.; Benson, I.C.: Studies on the elastic properties of human isolated aorta. J. Clin. Invest. *16:*595–602 (1937).

Johnson, P.C.: Autoregulation of blood flow. Circ. Res. *14/15,* suppl. I: (1964).

McDonald, D.A.: Blood Flow in Arteries; 2nd ed. (Williams and Wilkins, Baltimore 1974).

Whittaker, S.R.F.; Winton, F.R.: The apparent viscosity of blood flowing in the isolated hind limb of the dog, and its variation with corpuscular concentration. J. Physiol. (London) *78:*339–369 (1933).

17 Determinants of Arterial Pulse Pressure

The physician is often faced with the problem of hypertension and is expected to interpret the significance of a change in systolic or diastolic pressure or both (hence, of pulse pressure).

The classic studies on the hemodynamic variables that affect these pressures were carried out by Wiggers on an artificial circulation machine. Although the techniques he used are outdated, the essential findings of Wiggers still hold with newer stroke pumps and pressure transducers. In such a model the following variables can be altered *independently* of each other: stroke volume, rate/min, TPR, and arterial distensibility.

Stroke Volume

Increasing the stroke volume (keeping rate, TPR and arterial distensibility constant) causes a rise of both systolic and diastolic pressures, but the rise of systolic is more than that of the diastolic. Hence, pulse pressure is increased (fig. 122). Mean pressure is obviously elevated since output is increased while TPR is held constant. The reverse occurs when the stroke volume is decreased.

Stroke Frequency

Increasing rate, keeping *stroke volume,* TPR, and arterial distensibility constant, causes an increase in both the systolic and diastolic pressures, but the rise of diastolic pressure is more than that of the systolic (fig. 123). Hence, pulse pressure is reduced. Again the mean pressure rises because output is increased without a change in TPR. The reverse occurs when rate is reduced (other factors remaining constant).

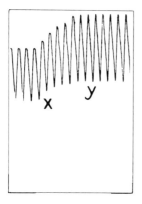

Fig. 122. Effect of increasing stroke volume (rate, TPR and elasticity constant). [Re-
produced, with permission, from Wiggers, C.J.:
Physiology in Health and Disease; 5th ed. (Lea
and Febiger, Philadelphia 1949).]

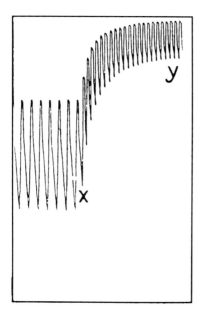

Fig. 123. Effect of increasing heart
rate (stroke volume, TPR and elasticity
constant). [Reproduced, with permission, from Wiggers, C.J.: Physiology in
Health and Disease. 5th ed. (Lea and
Febiger, Philadelphia 1949).]

Increasing rate, while keeping *output*, TPR, and arterial distensibility constant, causes a fall of systolic and a rise of diastolic pressures to
the same extent. Pulse pressure decreases markedly, but the mean arterial pressure remains constant since output and TPR remain unchanged
(fig. 124). Note that increasing the rate, while keeping *output constant*
reduces the stroke volume to an equal extent. The reverse occurs when
rate is reduced, while output and other factors remain constant.

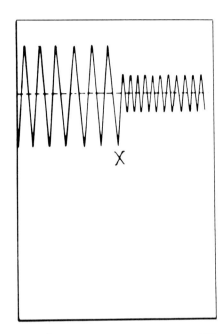

Fig. 124. Effect of increasing rate, keeping output, TPR and elasticity constant.

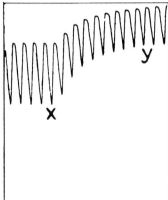

Fig. 125. Effect of increasing TPR (other factors constant). [Reproduced with permission, from Wiggers, C.J.: Physiology in Health and Disease; 5th ed. (Lea and Febiger, Philadelphia 1949).]

TPR

Increasing the TPR, while keeping stroke volume, rate, and arterial distensibility constant, causes a rise of both systolic and diastolic pressures, but the rise of diastolic is greater than that of systolic. Hence, pulse pressure is reduced (fig. 125). Mean arterial pressure obviously rises. The reverse occurs when TPR is reduced.

Arterial Distensibility (Compliance)

In the body this refers to the distensibility of the central arteries (aorta and its branches—windkessel).

Decreasing the arterial distensibility (increased rigidity or stiffness), keeping rate, stroke volume and TPR constant, causes a rise of systolic and a fall of diastolic pressures. Hence, pulse pressure rises markedly (fig. 126). The fall in diastolic is usually to the same extent as the rise in systolic, hence mean arterial pressure tends to remain unchanged (some German authors claim that mean arterial pressure rises).

The explanation of these results is based on the dynamic *pressure-volume relationship* of the *rubber tube* that Wiggers [1949] used in his model to represent the large central arteries. Likewise, in the body the dynamic compliance of the aorta and its branches plays a crucial role in determining the changes in pulse pressure. The distensibility of the rubber tube used in the model is somewhat like the curve shown in figure 127. Note that each stroke volume is accommodated by (a) driving some fluid from the elastic tube into the peripheral small vessels (systolic run-off) and (b) to a large extent is stored in the distensible tubes (stored volume or diastolic run-off). If we assume that the stored volume during ejection (Δv) is a constant proportion of the stroke volume and the duration of ejection is constant, we may use Δv to indicate the stroke vol-

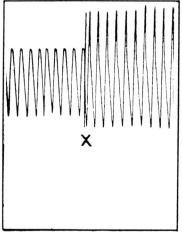

Fig. 126. Effect of decreasing compliance of rubber tube (other factors constant). [Reproduced, with permission, from Wiggers, C.J.: Physiology in Health and Disease; 5th ed. (Lea and Febiger, Philadelphia, 1949).]

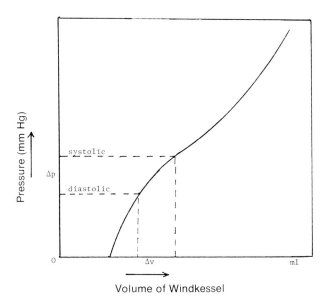

Fig. 127. The distensibility curve of the rubber tube believed to be used in the Wiggers' artificial circulation machine. Δv is taken to represent the stroke volume of the pump and Δp is the pulse pressure.

ume. Therefore, in young adults at rest with a normal stroke volume of about 75 ml, there is a normal Δp of about 40 mmHg (pulse pressure) (fig. 127).

When the stroke volume is increased with constant heart rate and duration of ejection, the output is increased and the pressure moves up on the distensibility curve (fig. 128). Increased stroke volume (Δv') produces an increase in pulse pressure (Δp') by increasing the systolic more than it does the diastolic. Obviously, mean pressure increases.

Increasing heart rate, while keeping stroke volume and other factors constant, also increases output and mean arterial pressure, thus shifting the pressure to the upper part of the curve. The results usually indicate a decline in pulse pressure, which depends on the shape of the curve at that particular arterial volume (fig. 129). On the other hand, increasing heart rate with a *constant output* decreases the stroke volume proportionately, causing a decline in pulse pressure but no change in mean arterial pressure (fig. 130).

A clinically important study is the effect of increasing TPR on the pulse pressure, other factors remaining the same. In the model, one

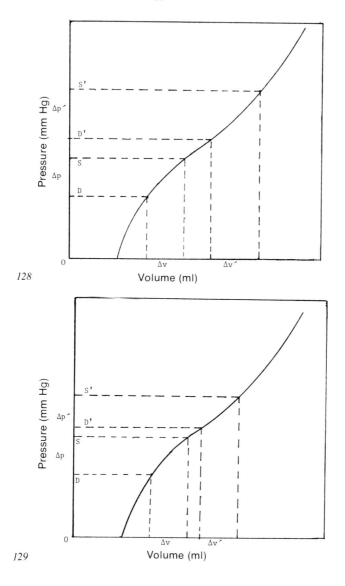

128

129

Fig. 128. The effect of increasing the stroke volume of the pump ($\Delta v'$) on pulse pressure. Mean pressure must increase (increased output), thus the new steady state must be up the curve. From the shape of the distensibility curve, it is obvious why the pulse pressure increases.

Fig. 129. The effect of increasing heart rate with constant stroke volume on pulse pressure. Output must increase and therefore the new steady state should be up the curve. From the shape of the curve it is clear why the pulse pressure decreases.

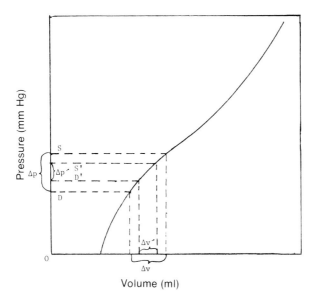

Fig. 130. The effect of increasing heart rate with constant output on pulse pressure. Mean pressure remains unchanged. Stroke volume decreases in proportion to the increase in heart rate; thus pulse pressure decreases.

usually observes a decrease in pulse pressure associated with an increase in mean arterial pressure (fig. 131). The effect on pulse pressure clearly depends on the shape of the pressure-volume curve. Clinical teaching emphasizes that TPR primarily affects diastolic pressure with little effect on systolic. This erroneous concept seems to have been derived from studies on the artificial circulation machine in which the characteristic distensibility of the rubber tube plays a crucial role in determining the results. If the aortic distensibility curve is *concave to the pressure axis* (as the work of Hallock and Benson suggest—see fig. 118, Ch. 15) then increased TPR will increase systolic pressure to a greater extent than the diastolic, a result contrary to clinical teaching!

The effect of decreased distensibility of the central arteries (eg, arteriosclerosis of the aorta) is to make the curve steeper. As a result systolic pressure goes up and diastolic falls (fig. 132) with little change in mean arterial pressure.

If several of these variables change simultaneously, one cannot predict the results because they would depend on the degree and direction

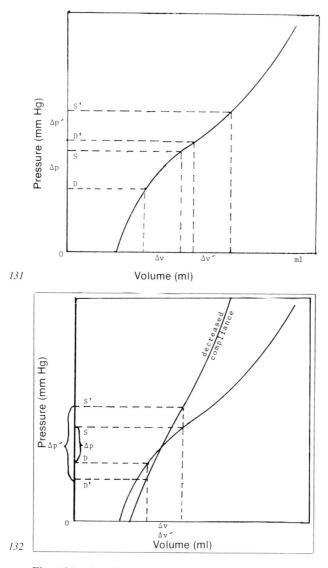

131

132

Fig. 131. The effect of increasing TPR on pulse pressure. Mean pressure must increase and the new steady state must be up the curve. From the shape of the curve, it is evident that the pulse pressure must decrease. If the shape was different, an opposite result would be obtained (see text).

Fig. 132. The effect of increasing the rigidity of the rubber tube (decreased compliance) on pulse pressure. Mean pressure is unchanged but the compliance curve is steeper. From the shape of the new curve, it is clear that systolic pressure must rise and diastolic pressure must fall, thereby markedly increasing the pulse pressure.

Table 3. Summary (as analyzed on circulation machine)

	Increased	Decreased
Stroke volume	systolic ++) diastolic +) p.p. +	systolic −−) diastolic −) p.p. −
Stroke frequency		
Constant SV and other factors	systolic +) diastolic ++) p.p. −	systolic −) diastolic −−) p.p. +
Constant output and other factors	systolic −) diastolic +) p.p. − −	systolic +) diastolic −) p.p. ++
TPR	systolic +) diastolic ++) p.p. −	systolic −) diastolic −−) p.p. +
Aortic compliance	systolic −) diastolic +) p.p. − −	systolic +) diastolic −) p.p. ++

of change of each variable. A frequent occurrence in clinical practice is an increase in TPR with decreased aortic compliance as is seen in certain cases of hypertension. This causes increased systolic pressure with normal or elevated diastolic (eg, 190/110 mmHg). Pulse pressure is usually increased.

Simple arteriosclerosis of the aorta (without any change in TPR) causes high stystolic and low diastolic without a significant rise of mean pressure. Clinical experience has shown that in these cases cardiovascular complications are less likely than in increased TPR with high *mean* and diastolic arterial pressures that *continuously* stretch the arterial walls excessively. These arteries tend to undergo pathologic changes that favor hemorrhage or thrombosis (the most serious being in the cerebral and coronary vessels). These studies are summarized in table 3.

Note: These results are critically dependent on the *distensibility curve of the rubber elastic reservoir* that corresponds to the arterial system in the living animal.

Reference

Wiggers, C.J.: Physiology in Health and Disease; 5th ed.; pp. 687–691 (Lea and Febiger, Philadelphia 1949).

18 Systemic Microcirculation and Lymph

The homeostatic functions of circulation and respiration are achieved in the systemic capillaries. There is a continual exchange of water, solutes, gases, and heat across the capillary walls to maintain the cellular steady state. Hence, capillary physiology is of vital importance to cellular function and life.

Structural Features of Capillaries

Capillary structure is adapted to its function in the form of a large surface area for exchange and a short distance for transit. Total systemic capillary surface area is estimated to be about 6000 m² and the average thickness of the capillary walls is about 0.5 μm (5000 Å), except at the nucleus. Figure 133 shows a sketch of an electron micrograph of a muscle capillary (nonfenestrated). Note the numerous *vesicles* at the cell border and within the cytoplasm. This suggests a process of transport known as *pinocytosis* or cytopempsis, possibly for the transfer of large protein molecules. The endothelial cell junctions overlap, showing a narrow tortuous space about 60–100 Å wide. Probably this represents the "pores" postulated by physiologists on the basis of passage of different molecular species. The tortuous spaces are supposed to contain an acid mucopolysaccharide (formerly referred to as intercellular cement substance). Mitochondria are few and endoplasmic reticulum is sparse. An amorphous basement membrane (200–600Å thick), surrounds the endothelial cells.

In endocrine organs, the endothelial cells of capillaries have regions of very thin membranes, with little cytoplasm, that are believed to facilitate the passage of hormones. These vessels are sometimes referred to as fenestrated capillaries.

In the spleen, liver, bone marrow, and lymph nodes the capillaries have large diameters and show discrete gaps between the endothelial

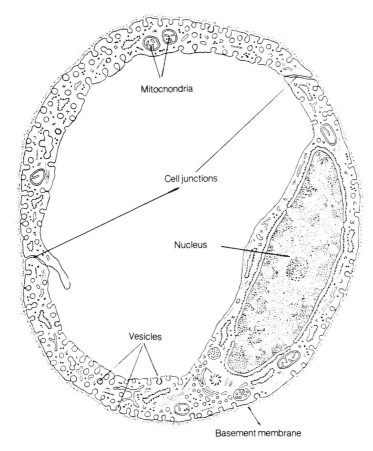

Mitochondria

Cell junctions

Nucleus

Vesicles

Basement membrane

Fig. 133. Idealized diagram of the endothelial cells (in cross section) of a continuous type of capillary, based on electron microscopic observations. [Labeled and reproduced from Lentz, T.L.: Cell Fine Structure. (W.B. Saunders Co., Philadelphia 1971).]

cells under light microscopy. These vessels are referred to as sinusoids. Relatively large gaps permit the passage of very large particles, even that of mature red cells in the bone marrow.

In the brain, the endothelial cells of the capillaries overlap markedly with tight junctions and without spaces suggestive of pores, except in the hypothalamic areas where several hormones are manufactured and released into the capillaries. The limited and selective permeability of cerebral capillaries has led to the concept of blood-brain barrier (BBB).

Capillary Circulation

This can be observed in the living state in the cat's mesentery, frog's web, rat's mesoappendix, dog's omentum, bat's wing, rabbit's ear, and so forth. The velocity of blood is about 1 mm/sec, due to the large cross-sectional area of *all* the functional capillaries.

$$\text{Capillary blood velocity} = \frac{\dot{Q} \text{ (cardiac output)}}{\text{Total x-area of all functional capillaries}}.$$

The average length of capillaries is about 1 mm. Therefore, it takes about 1 second for blood to pass through the capillaries.

Figure 134 shows a capillary network based mostly on observations of rat's mesoappendix. The arterioles $\rightarrow$ metarterioles $\rightarrow$ thoroughfare channels which are open at all times. The first two have smooth muscle. From these vessels arise the "true capillaries," which are guarded by a "precapillary sphincter" with smooth muscle. The true capillaries anastomose freely with each other and not all are open at any one moment, being regulated by the activity of the smooth muscle. Flow in the arterioles is slightly pulsatile with each heart beat, but in the true capillaries the pulsations practically disappear. Instead, the flow is on and *off* (intermittent) due to the contraction and relaxation of the precapillary sphincters (about 1–10 times/min). This phenomenon is referred to as vasomotion. In the venules, the velocity of flow increases, although the total volume flow is the same as in the capillaries. Why?

In some capillary beds there are arteriovenous anastomoses that have a layer of smooth muscle. They are not designed for exchange (nonnutrient) but seem to play a role in the distribution of blood by opening and closing. In skin vessels, the arteriovenous anastomoses alter blood flow to the skin, thereby affecting skin temperature and heat loss (exposure to heat opens up the arteriovenous anastomoses). In deep organs, their function is not entirely clear.

Transcapillary Exchange

The movement of water and solutes across capillaries involves at least two, or possibly three, processes:
1. *Diffusion*. This depends on thermodynamic activity of molecules (or concentration differences and solubility in different media). Osmo-

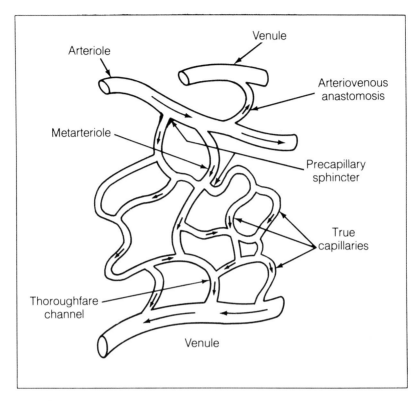

Fig. 134. Schematic diagram showing the arrangement of microvessels in a peripheral vascular bed. Blood flow in individual capillaries is intermittent (called vasomotion) due to the activity of precapillary sphincters. [Reproduced, with permission, from Little, R.C.: Physiology of the Heart and Circulation; 2nd ed. (Year Book Medical Publishers, Chicago 1981).]

sis is considered a special case of diffusion of solvent (H_2O) when some solute(s) cannot pass across a semipermeable membrane or passes with great difficulty (slowly).

2. *Filtration or Bulk Flow.* This depends on differences in mechanical pressure across the capillary wall with relative impermeability to some molecular species.

3. *Vesicular Transport.* Physiologists are not certain that vesicles represent the transport of large protein molecules across the capillary wall.

The capillary walls impose a barrier to diffusion and filtration in a manner not yet clearly defined. According to one view, there are "pores"

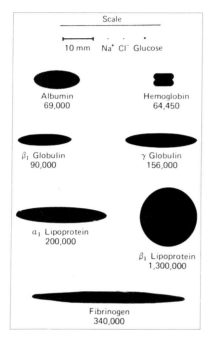

Fig. 135. The shape and molecular weights of some proteins in blood. [Reproduced, with permission, from Ganong, W.F.: Review of Medical Physiology, 10th ed. (Lange Medical Publications, Los Altos, CA. 1981).]

analogous to pores in artificial membranes (eg, cellophane) through which solutes that are lipid-insoluble pass. Pappenheimer used the Poiseuille equation and Fick's law of diffusion to calculate the pore radius (not diameter as stated in some texts) of the capillaries in the hindlimb of the cat. He reported a value of about 30–45Å. From net transfer rates of substances added to the perfusing blood and capillary surface area, he estimated that less than 0.2% of the total capillary surface area is available for the passage of H_2O and solutes that are lipid insoluble (eg, urea, glucose, amino acids, Na^+, K^+). This low value suggested that the "pores" are limited to the *intercellular* junctions. Later studies of Karnovsky with the use of horseradish peroxidase (estimated radius = 30Å) and histochemical localization with the EM supported the views of Pappenheimer.

It has been known for a long time that protein molecules pass across systemic capillaries with considerable difficulty. Their molecular equatorial diameter as well as their length seem to play a role (fig. 135). Albumin passes with greater ease than globulins or fibrinogen. This finding does not lend support to the idea that vesicular transport or pinocytosis is the basis for protein transfer, since this process is unlikely to distinguish molecular size whereas pores would. Protein concentration in the lymph from an organ is often taken to indicate the permeability of those capillaries to protein.

Plasma has 6–8 g protein/dl.
Lymph from the extremities has about 1% protein (fig. 136).
Lymph from the GI tract, kidneys, heart and lungs has about 3–4% protein.
Lymph from the liver has about 6% protein.
In the thoracic duct, where all of the above mix, there is about 3% protein.

With regard to blood gases (O_2 and CO_2), which are *lipid* soluble, there is evidence that the entire capillary surface is available for diffusion. The same applies to other lipid-soluble substances like fatty acids and cholesterol. Similarly, alcohol, which is lipid soluble, diffuses rapidly across the entire surface of the capillary endothelium. Anesthetic drugs like ether, chloroform, halothane, and so on, also diffuse readily by virtue of their high lipid solubility.

Leukocytes are capable of squeezing through the intercellular junctions and leave the capillary blood or return to the capillary (a process known as diapedesis). In acute inflammation due to infection, a large number of neutrophils passes through the capillary walls to fight the germs. Even an occasional RBC passes across the normal capillary wall.

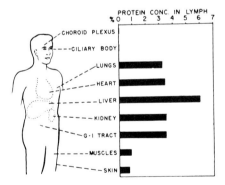

Fig. 136. Permeability of capillaries in various tissues and organs as judged by the concentration of protein in the lymphatics. The most permeable are the sinusoids of the liver. Plasma has about 7% protein. [Reproduced, with permission, from Rushmer, R.F.: Cardiovascular Dynamics; 2nd ed. (W.B. Saunders Co., Philadelphia 1961).]

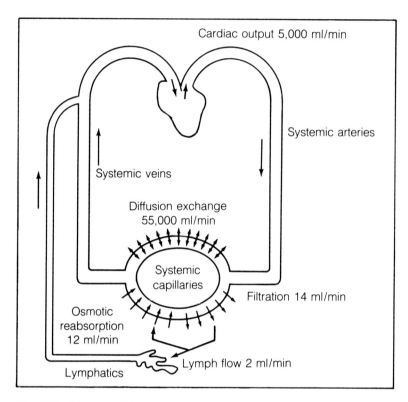

Fig. 137. Diagram of the systemic circulation to indicate the magnitude of capillary diffusional exchange, filtration-absorption and lymph flow. [Modified from Landis, E.M.; Pappenheimer, J.R.: Exchange of substances through the capillary walls; in Hamilton and Dow: Handbook of Physiology, Sect. 2: Circulation; vol. 2 (American Physiological Society 1963).]

Diffusion. Diffusional exchange across the capillary wall occurs *in both directions* at a very high rate. It is estimated that a total of about 55 liters of H_2O exchanges bidirectionally across the capillaries per minute. (This is *not a net* transfer) (fig. 137). Water is believed to diffuse through the endothelial cells. Various dissolved substances diffuse at a lower rate than water, depending on molecular size, lipid solubility, and so on. Nutrient substances that are continuously utilized by tissue cells, such as glucose, amino acids, lipids, and so on, develop a concentration gradient from plasma to interstitial fluid and, therefore, a *net* diffusion occurs in the direction of cells. On the other hand, metabolic end-products develop concentration gradients in the opposite direction, from cells to plasma. Thus, diffusion across capillaries is by far the most important mechanism for the supply of nutrients and the removal of waste products. With normal blood flow, equilibrium is reached quickly. Large quantities can be transported to meet metabolic needs. Capillary permeability to these substances is so high that it is *not* a limiting factor, but reduced blood flow can be.

Bulk Flow or Filtration-reabsorption. This is a much slower process. It is estimated that about 14 ml fluid per minute is filtered at the arterial end of all systemic capillaries; 12 ml/min is returned at the venous end (fig. 137). The difference—about 2 ml/min—enters the lymphatic capillaries. The amount of nutrients (eg, glucose) carried by this is very small compared to the net amount transferred by diffusion.

What then is the function of filtration-reabsorption in the capillaries? Its major function is to determine the volume of plasma in relation to the volume of interstitial fluid. Interstitial fluid serves as a reservoir to maintain the plasma volume, the constancy of which is critical for circulation. When plasma is lost, as in hemorrhage, diarrhea, vomiting, excess sweating, and so on, the total osmotic absorption across the capillaries is greater than the total filtration, thereby tending to restore plasma volume. The reverse occurs when there is an excess plasma volume (fluid accumulates in interstitial spaces—edema).

The regulation of this exchange is achieved by the interplay of two physical forces that are responsible for filtration-reabsorption. These forces were first identified by Starling and the concept is known as Starling's hypothesis. The forces are: (a) pressure inside the capillaries vs. the pressure outside, in the interstitial fluid (capillary transmural pressure); and (b) colloid osmotic pressure of the capillary blood *vs* the colloid osmotic pressure of the interstitial fluid. (Colloid osmotic pressure is often referred to as "oncotic" pressure.)

In a schematic presentation shown in figure 138, at the *arterial* end of the capillary the *net* or transmural capillary pressure is greater than the *net* or transmural oncotic pressure. As a result, water and solutes pass out of the capillary except for proteins. Hence, the term "filtration." At the venous end of the capillary, water and solutes are reabsorbed because (a) capillary pressure drops significantly due to frictional loss of energy and (b) oncotic pressure of plasma increases slightly because H_2O and solutes pass out at the arterial end leaving behind the

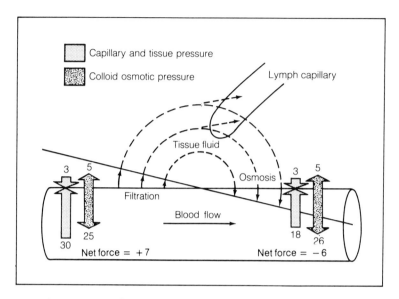

Fig. 138. Diagram of a capillary showing the forces involved in the exchange of fluid across the wall. (Magnitude of forces is approximate). The balance of forces is indicated by the oblique line.

colloids, which become slightly more concentrated. The balance of forces now is in favor of osmotic reabsorption of fluid. There is a *net* diffusion of solvent (H_2O) followed by a *net* diffusion of solutes into the capillary at the venous end. Note the small amount of protein in the interstitial fluid does not return to the plasma because the concentration gradient is in the opposite direction (from plasma to interstitial). It finds its way to the lymphatics, which are freely permeable to proteins.

Somewhere in between the two ends of the capillary, there is a point of *equilibrium* wherein the number of molecules filtering out equals the number of molecules returning by diffusion. Hence, there is no *net* transfer. Figure 138 is highly schematic. In a given capillary, the equilibrium point can be anywhere along the vessel. Some capillaries with high pressure may filter only; others with low pressure may absorb only.

Some authors prefer to express these forces in the form of an equation as follows:

$$\text{Net fluid transfer} = k\left[(P_c - P_{if}) - (\pi p - \pi_{if})\right]$$
(per unit time)

where k = filtration coefficient of the capillary walls; P_c = capillary pressure (dynamic + gravitational); P_{if} = pressure of interstitial fluid; π_p = plasma oncotic pressure; π_{if} = interstitial fluid oncotic pressure.

The *net* volume of fluid that is transferred across the capillaries (sometimes called *net* filtration) has been estimated experimentally in isolated perfused tissues and is expressed as a coefficient known as *capillary filtration coefficient*. Its unit is in milliliters fluid/minute per 100 g tissue per millimeter of mercury net pressure difference across the capillary walls, taking into account both the capillary and the oncotic pressures. The coefficient is an index of the characteristic *permeability* and *surface area* of functional capillaries in a particular organ or tissue.

It has been shown that capillary permeability of a given tissue is not changed by such physiologic conditions as arteriolar dilation and capillary distension, hypoxia, hypercapnea or increased [H⁺], but increases

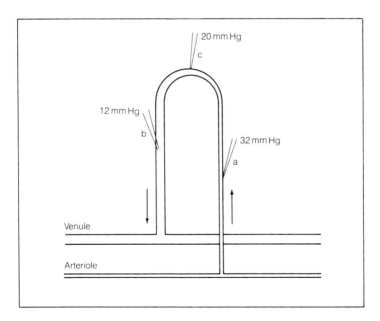

Fig. 139. Diagram of the hairpin loop in nail fold, showing the position of the pipette for measuring the pressures at various points. Note the drop of pressure along the loop. [Reproduced from Landis, E.M.: Micro-injection studies of capillary blood pressure in human skin. Heart *15:*209 (1930). Used with the permission of The Biochemical Society, London.]

when capillaries are injured by toxins or severe burns. Functional capillary surface area increases with the metabolic activity of the tissues by the opening up of closed capillaries.

Evidence in support of Starling's hypothesis came from direct measurements of capillary pressure by Landis [1963]. He punctured capillaries in frog's web and in the human nail bed with a micropipette connected to a manometer. Values of capillary pressure were consistent with Starling's hypothesis. Figure 139 shows the drop of pressure along the hairpin loop of the human nail bed.

The magnitude of pressures given in figure 138 represent the classic view. However, studies by Guyton and his co-workers [1975] have shown that the interstitial fluid pressure is negative, about -7 mmHg. Guyton estimates that the functional mean capillary pressure is lower than that shown in figure 138. (Mean capillary pressure = 17 mmHg; pressure at arterial end = 25 mmHg; and at the venous end = 9 mmHg). The problem of explaining the negativity of interstitial pressure remains open. According to Guyton [1973], it is due to the pumping activity of contractile cells around the lymph vessels which have valves to direct the flow. Guyton has not contested the essential features of the Starling hypothesis (only the magnitude of values has been altered). Not all investigators agree that interstitial fluid pressure is negative [Wiederhielm, 1968].

The protein concentration of fluid actually filtered at the arterial end has been estimated in an ingenious way. Lymph from the upper extremity at rest was obtained and its protein concentration was about 1%. Then, venous obstruction without arterial obstruction was produced by a blood pressure cuff set at a pressure of 40–50 mmHg. This raised the venous and capillary pressures sufficiently to prevent the reabsorption of fluid at the venous end of the capillaries. During this time, the lymph obtained had a protein concentration of only 0.2% (0.1%–0.5%). This low value represents the average protein concentration of the filtrate in the absence of reabsorption at the venous end.

If the filtrate contains 0.2% protein, how does one explain the 1% concentration of protein in the lymph from the extremities? *Filtered protein does not re-enter the capillary at the venous end but finds its way to the lymph capillaries,* which are freely permeable to protein. As water and other solutes re-enter the capillary at the venous end, the protein of the filtrate is concentrated from 0.2% to 1%. One of the important functions of lymphatics is to remove protein from the interstitial fluid, thereby keeping its protein concentration at a low level and preventing edema.

Control of Capillary Flow and Pressure

This is related to the activity of smooth muscle of the *precapillary sphincters* and of the *parent arterioles*. Arteriolar dilation or relaxation of sphincters will raise the capillary pressure and increase flow, and vice versa (fig. 140). The smooth muscle of these vessels is controlled by chemicals and nerves. (a) Chemicals—O_2 lack and metabolites from tissue cells are vasodilators. This is a very important feedback mechanism to adjust blood flow to the metabolic needs of tissues. Increased metabolism increases blood flow and vice versa, serving to maintain homeostasis. Also, hormones control these vessels. Catecholamines, angiotensin, and high concentrations of ADH cause the precapillary sphincters to contract. Bradykinin is a locally produced vasodilator during the secretion of the pancreas, sweat gland, and salivary glands. (b) Vasomotor nerves are postganglionic sympathetic. In general, they are adrener-

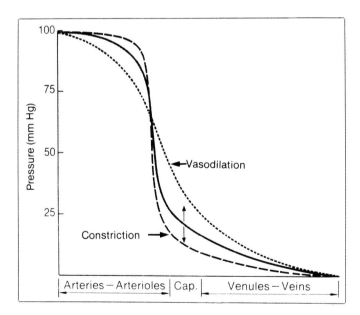

Fig. 140. Profile of pressure drop along a regional vascular bed (muscle) during rest (—); vasodilation (···); and vasoconstriction (---). [Reproduced with permission from Folkow, B.; Neil, E.: Circulation (Oxford University Press, Oxford 1971).]

gic constrictors (α-receptors), but some sympathetics to the vessels in skeletal muscle are *cholinergic dilators*.

Capillary Fragility

This refers to the breaking of the capillary endothelium or of the endothelial junctions, causing blood to escape into the interstitial spaces (not to be confused with capillary permeability to molecules). It is a remarkable fact that the 0.5μm thick capillary wall can withstand a mean transmural pressure of about 20 or 25 mmHg. Burton [1972] pointed out that this is related to the small radius (4μm) which causes a low tension in the wall. For a cylinder, T = P$\cdot$r (figs. 141 and 142). Tension in the wall of the aorta across its *entire thickness* is about 10,000 times greater than that of the capillaries. Therefore its wall is thicker to withstand this tension.

It is believed that the capillary wall is not very distensible. It can break by external trauma or by increasing the transmural pressure to high levels, eg, venous obstruction without arterial obstruction or excess negative pressure outside the capillaries (cupping produces hemorrhages). The capillary fragility test is done by obstructing veins in the upper extremity or by exerting known amounts of suction to the skin and counting the hemorrhagic spots. Fragility is high in the following conditions: (a) ascorbic acid or vitamin C deficiency (scurvy), (b) decreased number of blood platelets (purpura), (c) toxins of some bacteria

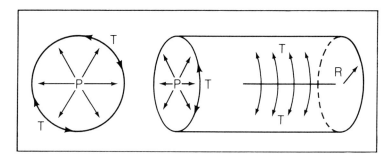

Fig. 141. Diagram illustrating the definition of tension in a vessel wall, namely the force pulling apart the wall from a longitudinal slit of a unit length, eg, dynes per centimeter length of slit. Tension = transmural pressure times radius of vessel (Laplace's equation). [Reproduced, with permission, from Burton, A.C.: Physiology and Biophysics of the Circulation; 2nd ed. (Year Book Medical Publishers, Chicago 1972).]

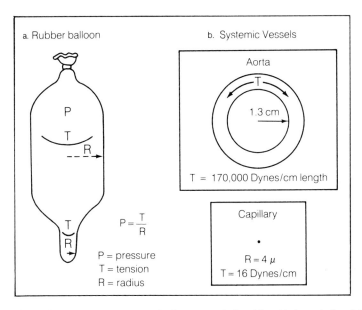

Fig. 142. Diagram showing the Laplace relationship. (a) is an inflated balloon the inside pressure of which is the same throughout but the wall tension is much greater in the distended portion than in the tip due to the difference in the radii of curvature. For a given pressure, a larger radius results in greater tension, T = P · r. (b) In the aorta, which has a much larger radius than a capillary, tension in the wall is about 10,000 times greater than that in an open capillary. Thus, the aortic wall must be much thicker to withstand the high tension without rupturing. [Reproduced, with permission, from Rushmer, R.F.: Cardiovascular Dynamics; 4th ed. (W.B. Saunders Co., Philadelphia 1976).]

or snake venoms, (d) allergic reactions, (e) some people with genetically "weak" capillary walls, and (f) diabetic microangiopathy.

Interstitial Fluid and Lymph

Interstitial fluid has the same composition as plasma, except for low protein and a somewhat different electrolyte concentration due to the Gibbs-Donnan equilibrium. There is more Na^+, K^+, Ca^{2+}, Mg^{2+}, and less Cl^-, HCO_3^-, $PO_4^{\equiv}$ in plasma. The reason is, plasma proteins are negatively charged (anions) at pH 7.4 and, therefore, attract cations and repel anions to a certain extent. The H^+ is an exception, being more concentrated in interstitial fluid, because it is derived from metabolic acids of tissue cells ($CO_2 + H_2O \rightarrow H_2CO_3 \rightarrow H^+ + HCO_3^-$). Interstitial fluid contains a *gel* made up of mucopolysaccharides, largely hyaluronic

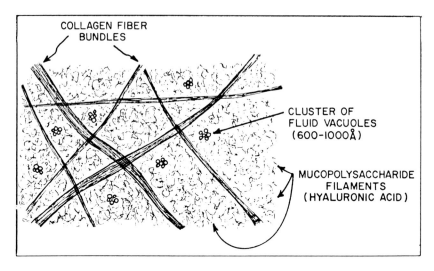

Fig. 143. Diagram showing the role of collagen and mucopolysaccharide filaments in tissue spaces for explaining interstitial fluid dynamics. [Reproduced, with permission, from Guyton, A.C.; Taylor, A.E.; Granger, H.J.: Circulatory Physiology II: Dynamics and Control of the Body Fluids. (W.B. Saunders Co., Philadelphia 1975).]

acid (fig. 143). This gel prevents interstitial fluid from moving freely, but it does not interfere significantly with the diffusion of molecules and metabolites. Also, the filtration and reabsorption of fluid along the length of capillaries are not interfered with.

All constituents of interstitial fluid, including the proteins, readily enter the lymph capillaries, which are as extensive as blood capillaries. Small and large lymph vessels have valves that permit unidirectional flow. Central lymph contains lymphocytes picked up from lymph nodes. During the absorption of food, intestinal lymph is rich in fats (chylomicrons), giving it a milky appearance (called chyle).

The driving force for lymph is the difference in pressure between interstitial fluid and the thoracic duct or the right lymphatic duct. The interstitial fluid pressure changes with the movement of tissue (both active and passive), arterial pulsations, compression from outside, gravity, and so on. In a resting extremity, there is practically no lymph flow. Increased formation of interstitial fluid → increased interstitial pressure → increased lymph flow. This may occur when capillary permeability to protein is increased or capillary pressure is increased. Some believe that there are contractile elements around the larger lymph vessels which promote lymph flow by reducing lymph pressure and cause the negative pressure in the interstitial fluid. The resistance of lymph nodes to the flow of lymph seems to be negligible.

Lymph is returned to the venous blood by way of the thoracic duct and the right lymphatic duct. Total daily volume in man is about 2–4 liters (or 2 ml/min).

Edema

Excess interstitial fluid is called *edema* which is a safety mechanism to maintain plasma volume. Mild degrees of edema are not clinically detectable and tissue pressure remains negative according to Guyton. Venous pressure must increase more than 15 mmHg before clinically detectable edema occurs—"safety factor." Advanced degrees of edema can be detected clinically by observing swelling of tissue, or by pressing with the finger and removing it to note a depression (pitting), or by noting a rapid gain in body weight.

The mechanism of edema involves either (a) an increase in the capillary transmural pressure or (b) a decrease in capillary transmural oncotic pressure (fig. 144). Both of these circumstances increase the pro-

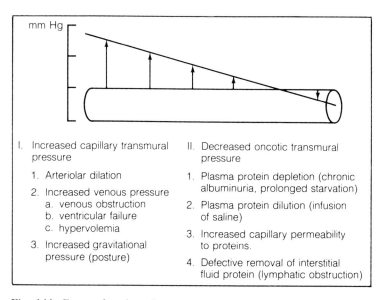

I. Increased capillary transmural pressure

  1. Arteriolar dilation

  2. Increased venous pressure
    a. venous obstruction
    b. ventricular failure
    c. hypervolemia

  3. Increased gravitational pressure (posture)

II. Decreased oncotic transmural pressure

  1. Plasma protein depletion (chronic albuminuria, prolonged starvation)

  2. Plasma protein dilution (infusion of saline)

  3. Increased capillary permeability to proteins.

  4. Defective removal of interstitial fluid protein (lymphatic obstruction)

Fig. 144. Factors favoring edema.

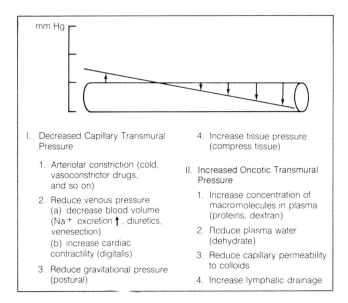

Fig. 145. Factors favoring absorption of edema.

duction of interstitial fluid and when lymphatic drainage is unable to keep up with the increased rate of production gross edema will result.

The treatment of edema is to reverse the situation, either by decreasing the capillary transmural pressure or by increasing the capillary transmural oncotic pressure (fig. 145). In practice, there are many ways of achieving these ends.

References

Guyton, A.C.; Taylor, A.E.; Granger, H.J.: Circulatory Physiology II: Dynamics and Control of the Body Fluids (W.B. Saunders, Philadelphia 1975).

Karnovsky, M.J.: The ultrastructural basis of capillary permeability studies with peroxidase as a tracer. J. Cell Biol. 35:213–236 (1967).

Landis, E.M.; Pappenheimer, J.R.: Exchange of substances through the capillary walls; in Hamilton and Dow: Handbook of Physiology; sect. 2; Circulation, vol. II; pp. 961–1034 (American Physiological Society, Washington, D.C. 1963).

Renkin, E.M.: Multiple pathways of capillary permeability. Circ. Res. 41:735–743 (1977).

Wiederhielm, C.A.: Dynamics of transcapillary fluid exchange. J. Gen. Physiol. 52:29s–63s (1968).

19 Control of Peripheral Vessels

Blood vessels, being distensible, may change their diameter either by external forces acting on the vessel wall (passive change) or by a change in the contractile state of the smooth muscle (active change).

Passive Changes

These are brought about by changes in the *transmural* pressure across the vessel wall (the difference between the inside and outside pressures). Note that the inside pressure has two components: (a) dynamic pressure resulting from pumping blood through the vessels, and (b) pressure due to the action of gravity on blood (ρgh) or gravitational pressure. The external pressure is represented by the tissue pressure which varies with many influences, eg, muscular contraction, volume of interstitial fluid, pressure of a capsule in an organ, intrapleural or intra-abdominal pressure, weight of body, negative pressure applied to skin, immersion of body in liquid (swimming), and so forth.

Active Changes

These depend on the contractile activity of vascular smooth muscle.

Vascular smooth muscle is subject to numerous influences which may be discussed under three headings: (a) neural, (b) chemical, and (c) mechanical.

Neural Control. In contrast to skeletal muscle, vascular smooth muscle is usually (but not always) innervated by two types of nerves. One type causes smooth muscle contraction—vasoconstrictor nerves; the other causes smooth muscle relaxation—vasodilator nerves.

The action of these nerves was first demonstrated by Claude Bernard [1851]. He cut the cervical sympathetic nerve on one side in a

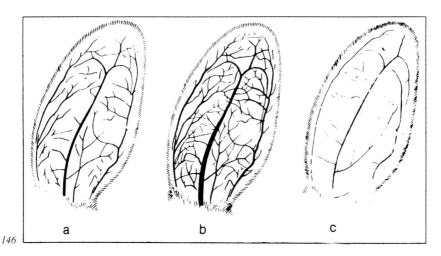

146

a   b   c

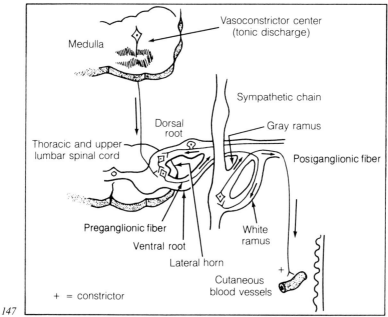

Medulla

Vasoconstrictor center
(tonic discharge)

Sympathetic chain

Dorsal
root

Gray ramus

Thoracic and upper
lumbar spinal cord

Postganglionic fiber

Preganglionic fiber

White
ramus

Ventral root

Lateral horn

Cutaneous
blood vessels

+ = constrictor

+

147

Fig. 146. Rabbit's ear vessels showing the influence of cervical sympathetic nerves. (a) Grossly visible vessels under resting conditions, (b) after cutting the cervical sympathetic nerves on that side, (c) during stimulation of the peripheral end of cut nerve on the same side. These experiments indicate that these nerves are vasoconstrictor and that they have a constant "tonic" constrictor action.

Fig. 147. Diagram of vasoconstrictor center in medulla and general pattern of vasoconstrictor nerve pathways to peripheral blood vessels.

rabbit, the ear on that side became more red, warmer, vessels were dilated and more vessels were visible compared to the opposite side (fig. 146). Later, Brown-Séquard stimulated the peripheral end of the cut cervical sympathetic nerve and observe the opposite effects (fig. 146).

These effects indicate that the nerves are vasoconstrictor to the ear vessels. The fact that cutting the nerve causes vasodilation shows that *normally* they carry impulses continuously to constrict these vessels, referred to as vasoconstrictor tone. Vasoconstrictor nerves belong to the *sympathetic* nervous system (fig. 147). Preganglionic neurons are located in the lateral horns of grey matter from $T_1 - L_2$. Their myelinated fibers travel in the ventral roots and go out in the white rami (placed laterally) to reach the sympathetic chain in the thorax and upper lumbar regions.

Some fibers synapse in the chain and the postganglionic (unmyelinated) fibers travel in the grey rami (placed medially) to join the somatic spinal nerves to go to blood vessels of the skin, the subcutaneous tissues, and skeletal muscles *all over* the body. Other fibers synapse in the chain or pass through without synapsing to go to *outlying sympathetic* ganglia where they synapse. Postganglionic fibers from these go to blood vessels of the viscera in the chest, the abdomen, the pelvic region, etc.

Postganglionic vasoconstrictor nerves liberate norepinephrine (adrenergic fibers) which acts on *α-receptors* located in the membrane of vascular smooth muscle. Some *adrenergic* sympathetic fibers to blood vessels of skeletal muscle are *vasodilators,* acting on *β*-receptors in some smooth muscle cells. Other sympathetic fibers to blood vessels of skeletal muscle are *cholinergic* vasodilators (can be blocked by atropine).

All peripheral arteries, arterioles, capillary sphincters, venules and veins in the body receive sympathetic vasoconstrictor fibers that have ''tonic'' activity. Tonic impulses may be traced to neurons in the medulla belonging to the reticular system. These cells constitute the *vasoconstrictor center or vasomotor center.* They are not a discrete group of cells that can be identified under the microscope (unlike the vagal nucleus) (fig. 147).

The vasomotor center is vital for circulation because its tonic activity maintains a proper balance between *total vascular capacity* and *blood volume,* thereby maintaining normal venous return, cardiac output, TPR, and blood pressure. If this tone is lost *completely,* generalized vasodilation of both resistance and capacitance vessels occurs, leading to a marked drop in cardiac output and blood pressure. Death may occur.

However, survival is not infrequent through the development of *intrinsic vascular tone* or activity of the lower spinal centers $(T_1 - L_2)$ and restoration of arterial pressure.

The normal tone of the vasoconstrictor center in the medulla is essentially determined by two influences:

1. *The normal P_{CO_2} and $[H^+]$ of arterial blood.* Normal arterial blood P_{CO_2} is about 40 torr (or mmHg) and pH is about 7.40. Both of these have an excitatory influence on the vasomotor center. Any *acute* decrease in P_{aCO_2} or $[H^+]$ below normal values reduces the vasomotor tone and blood pressure falls, eg, voluntary or psychic hyperventilation. Any acute increase in P_{aCO_2} or $[H^+]$ above normal increases the vasomotor tone and blood pressure rises, eg, breathing 5%–10% CO_2, asphyxia, hypoventilation, etc. The rise in blood pressure indicates that the action of arterial CO_2 on the tone of the vasomotor center is more powerful than the weak vasodilator action of CO_2 on the smooth muscle of peripheral vessels. There is some evidence to suggest that these changes in P_{aCO_2} and $[H^+]$ operate through changes in the $[H^+]$ of the cerebrospinal fluid. The functional value of the rise in blood pressure is not clear. It may be speculated that the rise is to increase cerebral blood flow and thereby wash away the CO_2 and H^+ in brain tissue, which is very sensitive to $[H^+]$. We shall note later that an increase in arterial P_{CO_2} is a powerful vasodilator of cerebral vessels, which may serve the same function (minimizing the accumulation of CO_2 and H^+ in brain tissue).

2. *Impulses normally arriving from the pressoreceptors* of the carotid sinuses and the aortic arch *inhibit* the discharge of the vasomotor center (for details refer to Ch. 22).

Vasodilator nerves, in contrast to vasoconstrictors, have a limited distribution. They are found not only in the parasympathetic nervous system but also in the sympathetics and in the somatic sensory nerves.

In the parasympathetic (cholinergic) system, the vasodilator nerves supply the blood vessels of the genitalia, the urinary bladder and the colon. The most striking effect is on the vessels of the erectile tissues. In the cranial portion of the parasympathetic, the nerves dilate the vessels of the salivary and lacrimal glands. It is not certain whether or not the vasodilation in these is exclusively the effect of A-Ch on the vascular smooth muscle because during secretion a vasodilator polypeptide,

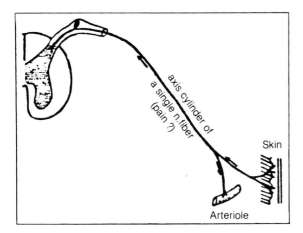

Labels in figure: axis cylinder of a single n. fiber (pain ?); Skin; Arteriole

Fig. 148. Diagram of nervous connections involved in axon reflex.

called bradykinin, is formed locally in these organs (probably both are involved).

We have already referred to the *sympathetic vasodilators* in the skeletal muscles. Here, some of the vasodilator fibers are adrenergic (NE), acting on the β_2-receptors of the smooth muscle cells, while others are cholinergic. Fibers of the cholinergic sympathetic dilator system arise in the cerebral motor cortex and pass through the hypothalamus and ventral medulla before reaching the sympathetic outflow in the spinal cord. They do not have tonic activity but are activated when the body is in danger (fight-or-flight situations). The value of the vasodilation in skeletal muscles is to increase blood flow in anticipation of the use of the muscles.

Vasodilators in the *somatic* nerves have a peculiar arrangement. Sensory fibers from the skin give off collaterals that go to the cutaneous vessels in the dermatomal distribution of the nerve. Irritants that stimulate certain cutaneous nerves (probably pain fibers) cause vasodilation in the area of distribution of the sensory nerve. This vasodilator response has been described as an axon reflex (fig. 148).

As to the question of a *general vasodilator center* in the medulla, most workers believe that there is no such center, contrary to some textbooks. Besides, such a center would not serve a useful purpose!

Vasomotor changes are detected by recording changes in blood flow, in the volume of a part or organ, and in the arterial pressure. In intact animals it is difficult to record blood flow accurately (volume and arterial pressure are easier to record).

Chemical Influences. Chemicals derived from tissues or organs may act *locally* or enter blood and act *diffusely* on distant vessels.

Local chemical mechanisms are most important in regulating blood flow to a given tissue according to its metabolic needs. Metabolic products are known to have a *powerful vasodilator action.* Thus, increased functional activity leads to more metabolites, local vasodilation, greater blood flow to the active tissue, and *vice versa.* This is an excellent self-regulating mechanism. The increased blood in an organ due to its functional activity is called *active hyperemia.* The term *hyperemia* or *congestion* refers to the increase in the *volume of blood* at any one moment and not to the increase in flow rate. Active hyperemia is best seen in skeletal muscles. Rhythmic contractions of skeletal muscles cause a marked increase in blood flow through the muscle (fig. 149). The hyperemia lasts for a long time after cessation of contractions. The metabolic products that dilate the vessels are not known with certainty. Suspected metabolites are CO_2, lactic acid, H^+, K^+, histamine, some

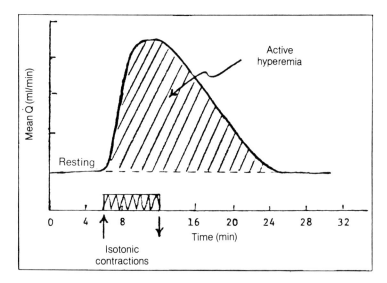

Fig. 149. Diagram of changes in muscle blood flow during and immediately after isotonic muscular exercise (active hyperemia).

prostaglandins and adenosine. No single metabolite has been shown to be *the* metabolite. Probably all of these contribute, perhaps some more than others. In addition to metabolites, there is some evidence that decreased P_{O_2} in the interstitial fluids of the active tissue, as a result of increased oxygen consumption or arterial hypoxemia, causes relaxation of vascular smooth muscle by *direct* action [Guyton 1981]. The role of oxygen is somewhat controversial because experimentally in vivo it is almost impossible to dissociate oxygen supply from tissue metabolism and metabolite production, the two are very intimately related. Whatever the true mechanism is, it is clear that blood flow is normally adjusted to the metabolic needs of the tissue.

The metabolic control of blood flow has been used to explain the constancy of blood flow when perfusion pressure is altered under steady metabolic states (autoregulation of blood flow). Increased perfusion pressure temporarily increases the blood flow which accelerates the removal of vasodilator metabolites, and the vessels constrict. The increased vascular resistance maintains the flow constant despite the increase in $P_1 - P_2$. The reverse occurs when $P_1 - P_2$ is reduced. This is the *metabolic* hypothesis to explain autoregulation of blood flow.

The metabolic control of blood flow is also noted when the oxygen supply of a tissue is inadequate for the needs of the tissue. This may occur when arterial P_{O_2} is reduced below its normal value of 95 mmHg (a condition called arterial hypoxemia). In moderate degrees of arterial hypoxemia the vessels dilate and increase the blood flow thereby maintaining the supply of oxygen. However, when the vasodilation reaches its limit, any further reduction in arterial P_{O_2} will result in insufficient O_2 supply and the tissue will suffer metabolic "injury" (hypoxia). The "injury" may be reversible or irreversible depending upon the severity and duration of tissue hypoxia. Another condition in which O_2 supply may be inadequate is severe anemia (deficiency of hemoglobin). Here, arterial blood P_{O_2} is normal but the transporting agent (Hb) is below normal. In severe anemia the peripheral vessels dilate and the TPR is lower than normal.

In some glands, like the salivary, sweat, and pancreas, during active secretion a substance called *bradykinin* has been found, which has a powerful vasodilator action *locally*. These glands liberate, during secretion, an enzyme called kallikrein. Kallikrein acts on α_2-globulins in the interstitial fluid to produce a *nonapeptide*, bradykinin, which acts locally as a vasodilator. This substance may be *partly* responsible for the increased blood flow to the salivary glands, pancreas, and sweat glands during their secretory activity.

Another type of hyperemia is *reactive hyperemia*. If one obstructs blood flow to a tissue for a period of time and then releases the obstruction, one observes an increase in blood flow above the control level for a short period of time (fig. 150). The increased flow and volume of blood in the organ is known as reactive hyperemia. Several factors play a role in causing the vasodilation: (a) accumulation of metabolites during the period of ischemia, (b) fall of transmural pressure in vessels distal to the obstruction causing the relaxation of smooth muscles (myogenic response), and (c) decreased interstitial fluid P_{O_2} during ischemia relaxes the smooth muscles. The intensity and duration of reactive hyperemia are much less than that of active hyperemia. Reactive hyperemia is looked upon as a payment of the metabolic debt to the tissue

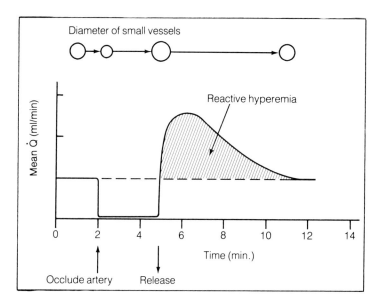

Fig. 150. Diagram illustrating changes in blood flow during an arterial occlusion and after release to a resting skeletal muscle. Under these conditions, reactive hyperemia is detectable only when the interferance with blood flow lasts a few minutes. Compare this with the duration of coronary occlusion (10–15 sec) in the pumping heart to produce reactive hyperemia in the myocardium. Remember that the metabolic rate of the heart is much greater than that of resting skeletal muscle.

incurred during the period of ischemia. The nerves to the blood vessels are not involved in the phenomenon. Reactive hyperemia is an everyday occurrence, eg, pressure points on skin, buttocks, and so on. It occurs after the use of vasoconstrictor drugs, eg, nasal vasoconstrictors. When the action of the drug wears off, there is more congestion.

A third type of hyperemia is *passive hyperemia or congestion*. This is produced by an obstruction to venous return or ventricular failure, causing rise of venous and capillary pressures, which in turn increases the volume of blood in an organ or tissue (hyperemia). Note that in this case the blood flow rate (Q) through the tissue is reduced, whereas in active and reactive hyperemia, blood flow rate is increased. If the venous obstruction is relieved, the passive hyperemia is followed by reactive hyperemia.

So far we have considered chemicals acting locally. Besides these there are hormones that circulate and act on *distant* blood vessels. Catecholamines from the adrenal medulla and the renin-angiotensin from the kidney are best known examples.

Mechanical Stretch. Bayliss [1902] suggested that the normal *transmural* pressure in small vessels contributes to the "tone" of smooth muscles. Recent studies support this concept. Increased transmural pressure causes smooth muscle *contraction* and vice versa. One hypothesis that explains autoregulation of blood flow is based on this smooth muscle response. It is postulated that an increase in perfusion pressure increases the vessel *wall tension* passively, causing smooth muscle contraction, which reduces the radius to restore wall tension in accordance with the Laplace equation (T (constant) = $P\uparrow r\downarrow$). The reduction in vascular diameter increases the vascular resistance to balance the increase in $P_1 - P_2$ and maintain $\dot{Q}$. This is the *myogenic hypothesis* to explain blood flow autoregulation.

The metabolic control and the mechanical influences on vascular smooth muscle play an important role in explaining certain phenomena. If a large artery is partially constricted, there is no decrease in blood flow under steady states until a *critical* constriction is reached, usually up to a decrease of 60%–80% of the cross-sectional area of the artery (critical stenosis). The mechanism seems to involve a dilation of distal small vessels due to the accumulation of metabolites, decreased P_{O_2} and a decrease in wall tension from the fall in pressure distal to the constric-

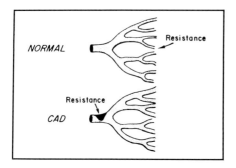

Fig. 151. Diagram showing effect of lumen narrowing in a large coronary artery (coronary artery disease, CAD) on diameter and resistance of arterioles supplied. Compensatory vasodilation (probably by metabolic mechanisms) maintains total vascular resistance and flow within normal limits up to a critical narrowing of about 60–80% decrease in cross-sectional area of the large artery. This is a protective mechanism to minimize myocardial ischemia in the early stages of coronary artery disease. [Reproduced, with permission, from Hood, W.B.: Pathophysiology of ischemic heart disease. Prog. Cardiovasc. Dis. 14:297 (1971).]

tion (fig. 151). The range of peripheral vasodilation from "normal" to maximum dilation reached at the critical stenosis of the parent vessel is an index of the "vascular reserve" of a given vascular bed.

References

Bayliss, W.M.: On the local reactions of the arterial wall to changes of internal pressure. J. Physiol. (London) *28:*220–231 (1902).

Folkow, B.: Nervous control of the blood vessels. Physiol. Rev. *35:*629–663 (1955).

Haddy, F.J.; Scott, J.B.: Metabolically linked vasoactive chemicals in local regulation of blood flow. Physiol. Rev. *48:*688–707 (1968).

Haddy, F.J.; Scott, J.B.: Metabolic factors in peripheral circulatory regulation. Fed. Proc. *34:*2006–2011 (1975).

20 Venous Pressure and Venous Return

Central venous (or right atrial) pressure is an important index of overall cardiac performance. This pressure depends on right atrial inflow and outflow of blood in relation to the distensibility of the atrium. As already pointed out, *alterations in venous inflow* into the ventricle are mainly responsible for changes in cardiac output in health and in disease. However, in some conditions changes in ventricular contractility (positive or negative inotropic effects) may be responsible for changes in cardiac output. During these *nonsteady* (transient) states there is a *rapid adjustment* in the circulation so that a new steady state is reached when cardiac output and venous inflow are again equalized. In *acute* heart failure cardiac output is less than inflow for a *short* period of time, after which a new steady state is established at a lower output characterized by a greater residual blood volume in the ventricle and elevated atrial and venous pressures.

The venous system is characterized by its large capacity (volume) and great distensibility compared to other parts of the vascular tree. The *compliance* (or capacitance) of veins is about 20 times greater than that of the systemic arteries. Because of this, about 50% of all the blood is found in the systemic veins at any one moment, and veins are described as *capacitance* vessels. If blood (or fluid) is added to or removed from the cardiovascular system, the adjustment in the system's volume is *chiefly* (but not entirely) on the venous side. However, the change in venous pressure is slight because of the large capacity and the great distensibility of veins.

If the heart suddenly stops beating (output = zero), pressure in the systemic vessels and the right heart equalizes at about 7 mmHg. This equilibrium pressure is called by Guyton [1973] the mean systemic filling pressure and is a function of the systemic blood volume in relation to the mean distensibility or compliance of all systemic vessels (fig. 152). Under these circumstances, venous inflow is *zero*. In such a model,

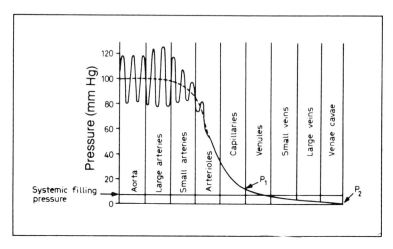

Fig. 152. Diagram showing the perfusion pressure ($P_1 - P_2$) responsible for venous blood flow and the difference between end-capillary pressure (P_1) in the dynamic state of the circulation and mean systemic filling pressure of Guyton, when the circulation is suddenly arrested. [Reproduced from Badeer, H.S.: Cardiac output and venous return as interdependent and independent variables. Cardiology *67:* 65 (1981) as modified from Guyton, A.C. Function of the Human Body; 4th ed. (W.B. Saunders, Philadelphia 1974).]

if the heart again starts beating at a normal rate and pumps out about 5 liters/min, the mean arterial pressure rises to about 100 mmHg and right atrial pressure falls to about 2 mmHg when a steady state is reached. During this period, some blood is transferred from the venous to the arterial system after inducing a similar shift in the pulmonary circuit. Note that the rise in arterial pressure (from 7 to 100 mmHg) is much more than the fall in venous pressure (from 7 to 2 mmHg), although the volume added to the arteries is almost equal to the volume subtracted from the veins (assuming not much is added to the pulmonary circuit). The reason for this is the difference in compliance or capacitance of the arterial and venous systems. Under these circumstances, the venous inflow to the heart will be about 5 liters/min. From such considerations, it is clear that the *basic factor responsible for venous return* or *inflow into the right ventricle is the energy imparted to the blood by the contraction of the left ventricle, which creates energy gradients* (chiefly in the form of pressure) *causing flow in the systemic circuit* (similarly, the right ventricular contraction is responsible for the venous inflow into the

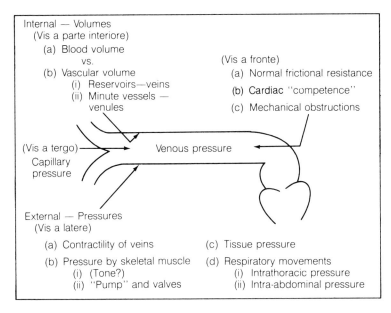

Fig. 153. Factors concerned in the maintenance of venous pressure. [Reproduced, with permission, from Landis, E.M.; Hortenstine, J.C.: Functional significance of venous blood pressure. Physiol. Rev. *30:*1 (1950).]

left ventricle). Later, it will be noted that, although ventricular contraction is basically responsible for venous return, *alterations* in venous return in health and in disease are more often due to *changes in peripheral circulation* than to changes in cardiac pumping. For this reason, Guyton refers to the heart as playing a "permissive" role in venous return. However, there are a few situations in which the contractile activity of the heart is primarily responsible for changes in venous inflow (eg, myocardial infarction).

A more careful analysis indicates that the pressure gradient responsible for venous return into the right heart is the difference between the pressure at the end of all systemic capillaries (P_1) and the mean right ventricular pressure during the three filling phases (P_2) (fig. 152). Normally, P_1 is about 15 mmHg and P_2 is about 2 mmHg. The end-capillary pressure is often referred to as the *vis a tergo* (force from behind) and plays an important role in altering the venous return in health and in disease. Other things being equal, *increase in this pressure will increase venous return and vice versa.* Some authors refer to P_1–P_2 as the *vis a*

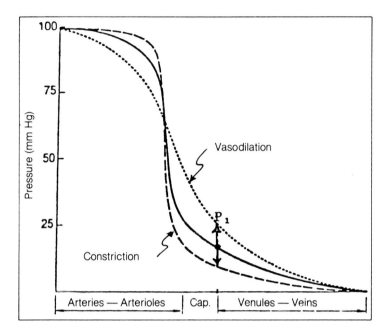

Fig. 154. Profile of pressure drop along a regional vascular bed (muscle) during rest (—); arteriolar dilation (···); and constriction (---). Note the effect of vasodilation and vasoconstriction on end-capillary pressure (P_1 or *vis a tergo*) that drives the blood through the venous system. [Modified from Folkow, B.; Neil, E.: Circulation. (Oxford University Press, Oxford 1971).]

tergo, but we will use the former definition (fig. 153). As a general rule, *moderate, arteriolar dilation tends to increase vis a tergo and venous return and vice versa* (fig. 154). This is the most frequent cause of changes in venous return in health and in disease. Occasionally, P_2 is increased and reduces the venous inflow, eg, fluid in the pericardial sac. Changes in venous return do result from changes in the *contractile activity* of the ventricles but are rather less frequent (eg, myocardial infarction, paroxysmal tachycardia or positive inotropic interventions).

The *vis a tergo* is confronted by the total resistance of the systemic veins and the right ventricular pressure during the filling phases of diastole. The "opposing force" is sometimes called the *vis a fronte* (fig. 153). P_2 is determined by right ventricular blood volume during filling and right ventricular diastolic compliance. Chronic right heart failure increases the ventricular blood volume and decreases the right ventricular compliance (due to hypertrophy) so that P_2 rises significantly. The total resistance of the veins to blood flow can be altered by active or passive changes in the diameter and shape of these vessels, which can have both temporary and sustained effects on venous return. Note that slight changes in P_1-P_2 can have marked effects on the venous return of blood because the total resistance of veins is rather low.

Because veins have thin walls and low pressure, they tend to collapse readily by pressure from outside. Consequently, many veins in vivo are not circular but are flattened (offer more resistance than circular). Venous obstruction (eg, thrombosis, tumors) increases the *vis a fronte* and venous pressure distal to the obstruction rises to maintain flow. This is the so-called "back pressure" effect. If venous obstruction is not extensive over a wide area, returning blood is directed to other veins in the region. The rise of capillary pressure tends to cause edema in the area drained by the obstructed vein.

The smooth muscle of veins and venules is subject to sympathetic vasoconstrictor control (α-adrenergic) and is subject to the normal vasomotor "tone." Changes in this tone can alter their capacitance and resistance to flow and affect venous return.

Changes in blood volume also alter venous return. An increase in blood volume (hypervolemia) increases venous return and cardiac output by increasing the mean systemic filling pressure and thereby increasing the *vis a tergo*. In hypovolemia the reverse is true (fig. 153).

Auxiliary Factors Affecting Venous Return

Contraction of Skeletal Muscles. Contraction squeezes on all vessels in the active muscles. The collapse of capillaries and veins is more than that of the arteries. Muscle contraction *momentarily* increases venous outflow from and impedes arterial inflow to the muscle and vice versa. If contraction is maintained *tetanically,* blood flow through muscle is reduced but does not stop completely. However, *intermittent* contraction of muscles, which is more common in everyday life, increases the overall venous return to the heart and increases muscle blood flow markedly by local vasodilation (exercise decreases TPR). During muscle relaxation, the valves in veins prevent the reflux of blood but this is not absolutely necessary because capillary pressure is higher than venous and, hence, flow is towards lower pressure (right heart). This effect of contracting muscle in aiding venous return and circulation is very important

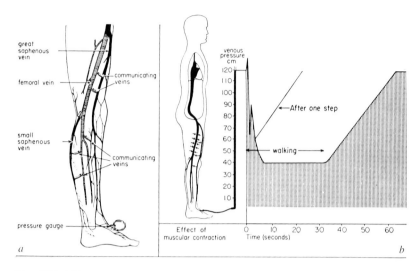

Fig. 155. Effect of muscle contraction on venous pressure in the lower extremity. (a) Communications between superficial and deep veins and location of pressure gauge in a vein on the dorsum of the foot. (b) Effect of taking a step or walking, in reducing venous pressure in foot. Contraction of muscles drives blood towards the heart. During relaxation, reflux of venous blood is prevented by the venous valves. As a result, peripheral veins are partially emptied and pressure is kept low. [Reproduced, with permission, from Rushmer, R.F.: Cardiovascular Dynamics; 4th ed. (W.B. Saunders, Philadelphia 1976).]

functionally and is referred to as the muscle pump. In the extremities, intermittent muscle contractions tend to reduce venous pressure by emptying the veins (fig. 155).

Paralyzed patients and patients confined to bed for some time have a reduced cardiac output (reduced muscle activity). Also, this is a problem in the space capsule where muscle activity tends to be limited.

Pressure in Thorax. Pressure in the chest outside the lungs is below atmospheric, fluctuating between -4 and -8 mmHg (negative). It is exerted on all intrathoracic structures, including the veins and atria. It tends to increase the in vivo transmural pressure of large veins and atria and reduces the P_2. This favors venous return to the heart. In the abdomen the pressure is slightly above atmospheric (positive), tending to collapse abdominal veins and impede flow from the lower extremities.

Thoracotomy (opening the chest) abolishes the negative pressure. This causes the heart and the intrathoracic veins to collapse partially

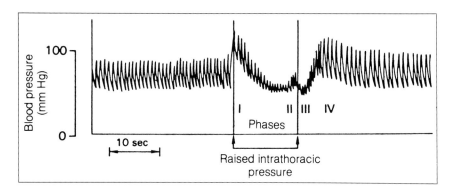

Fig. 156. Blood pressure response to Valsalva test in normal young adult. Intrathoracic pressure, 40 mmHg. Note falling arterial pressure in phase II. The pressure overshoot in phase IV indicates baroreceptor reflexes are operating in phase II, during which peripheral vasoconstriction occurs. Overshoot is partly due to increased output of venous blood, which had accumulated during phase II. [Reproduced, with permission, from Smith, J.J.; Kampine, J.P.: Circulatory Physiology—The Essentials. (Williams and Wilkins, Baltimore 1980).]

and impede venous inflow. As a result, cardiac output is somewhat decreased.

During normal inspiration, intrathoracic pressure falls and intraabdominal pressure rises. Both of these favor venous return *temporarily*. During expiration, the reverse occurs. The *net* influence of normal respiration on venous inflow over a period of time seems to be a slight increase.

Similarly, increase in rate and depth of breathing in a resting person (hyperventilation) increases venous return slightly (reduces arterial blood CO_2 and decreases the tone of the vasoconstrictor center in the medulla, causing vasodilation, leading to increased *vis a tergo*).

If the intrathoracic pressure is made positive by *forced expiration with the glottis closed* (Valsalva maneuver), the intrathoracic veins and the heart are severely compressed, P_2 rises, and venous inflow is impeded. Peripheral venous pressure rises, neck veins are engorged, and the face and eyes are congested. Cardiac output and blood pressure drop (fig. 156, Phase II). Carotid sinus reflex causes peripheral vasoconstriction and cardioacceleration. At the same time increased arterial P_{CO_2} and $[H^+]$ from breath holding stimulates the vasomotor center, to cause vasoconstriction. Upon release of the high intrathoracic pressure, blood

Table 4. Alterations in cardiac output

Conditions	Increased cardiac output (venous return)	Decreased cardiac output (venous return)
Physiologic	1. Intake of food and drink 2. Exposure to heat 3. Shivering from cold 4. Anxiety with increased metabolism 5. Pregnancy 6. Muscular exercise	1. Standing erect 2. Valsalva's maneuver 3. Muscular inactivity (anesthesia)
Pathologic	1. Severe anemia 2. Fever 3. Hyperthyroidism 4. A-V fistula 5. Beriberi 6. Hypervolemia 7. Patent ductus arteriosus (only to left heart)	1. Hemorrhage and shock 2. Acute myocardial infarction ⎫ pump 3. Congestive heart failure ⎭ failure 4. Severe dehydration and hypovolemia 5. Thoracotomy 6. Cardiac tamponade (fluid in pericardial sac) 7. Paroxysmal tachycardia 8. Extensive paralysis

Note: Analysis of the above indicates that changes in venous return and cardiac output are more frequently dependent on changes in *peripheral circulation* or *blood volume* than on cardiac pumping function (permissive role of heart in determining venous return).

pressure overshoots, which results in bradycardia due to the temporary increase in cardiac output while the peripheral vessels are constricted (fig. 155, Phase IV). Changes in venous return and cardiac output in health and disease are summarized in table 4. Explain the mechanism in each case on the basis of factors discussed in the text.

References

Badeer, H.S.: Cardiac output and venous return as interdependent and independent variables. Cardiology (Basel) *67:*65–72 (1981).

Guyton, A.C.; Jones, C.E.; Coleman, T.G.: Circulatory Physiology: Cardiac Output and Its Regulation; 2nd ed. (W.B. Saunders, Philadelphia 1973).

Landis, E.M.; Hortenstine, J.C.: Functional significance of venous blood pressure. Physiol. Rev. *30:*1–30 (1950).

21 Effect of Gravity on Circulation

This is frequently a misunderstood subject because it is thought that in the upright position gravity opposes and hinders the flow of arterial blood to the parts above the level of the heart and likewise opposes the return of venous blood from the lower parts of the body. This concept is incorrect because the vascular system is a *closed* circuit. The arteries to the head connect with the returning veins by way of the capillaries forming a kind of an inverted-U system of tubes. In this type of arrangement the *gravitational* (hydrostatic) component of pressure in the arteries is counterbalanced by the *gravitational* pressure of blood in the veins, similar to the situation that occurs in a siphon. Hence, gravity *per se* poses no impediment to the arterial flow to upper parts of the body.

Similarly, gravity does not hinder venous return from the lower parts of the body in the erect position because the *gravitational* pressure of blood in veins is balanced by the gravitational pressure of blood in the arteries to lower parts of the body. Thus, in a *closed* system gravity neither favors downward flow nor hinders upward flow. As Burton [1972] put it so aptly, "It is no harder, in the circulation, for the blood to flow uphill than downhill." These principles are illustrated in models shown in figures 157 and 158.

However, there is an important factor to be considered: the *tubes or vessels must be perfectly rigid* if gravity is to be without effect on circulation. This is illustrated in the model shown in figure 159. Positional changes of such a system will alter transmural pressures at different points due to changes in gravitational (hydrostatic) pressure in the fluid column, but this will not affect the tube *diameter* because of the rigidity of the tube. Hence, the input of fluid to the pump, the output of the pump, the resistance to flow, and the total *energy* gradients (see Bernoulli's equation) will not be altered. The energy expenditure of the pump will remain unchanged. Thus, *if our blood vessels were perfectly*

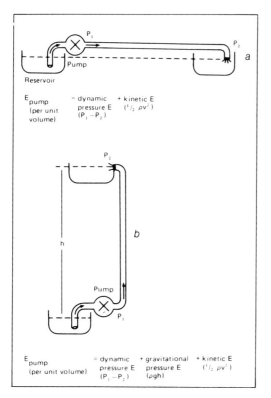

Fig. 157. The effect of gravity on the work of a pump driving liquid through a straight rigid tube. (a) The outlet of the tube is at the same level as that of the liquid in the reservoir. The energy of the pump is expended only in developing dynamic pressure differences that force fluid through the resistance of the tube and also to impart kinetic energy to the liquid. Note that in the figure and the equation, for simplicity the work of the pump in moving the liquid from the reservoir through the tube on the left side of the pump has been omitted. (b) The same tube is placed vertically and the same flow rate is maintained. The pump now uses more energy. The excess energy is due to the extra work done in overcoming gravitational (hydrostatic) pressure energy (ρgh), which is converted into gravitational potential energy (ρgh) of the liquid at the aperture of the tube. [Reproduced from Badeer, H.S.; Rietz, R.R.: Vascular hemodynamics: deep-rooted misconceptions and misnomers. Cardiology *64:*197 (1979).]

rigid nonleaky tubes, postural changes would have absolutely no effect on circulation.

However, this is not the case; very significant changes occur in the circulation with changes in posture. The essential factor is the *distensibility* of vessels. Postural changes alter the *transmural* pressure at dif-

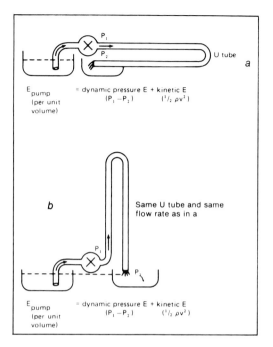

E_{pump} (per unit volume) = dynamic pressure E + kinetic E
$(P_1 - P_2)$ $(\frac{1}{2} \rho v^2)$

U tube *a*

b

Same U tube and same flow rate as in a

E_{pump} (per unit volume) = dynamic pressure E + kinetic E
$(P_1 - P_2)$ $(\frac{1}{2} \rho v^2)$

Fig. 158. The effect of gravity on the work of a pump driving liquid through a rigid U tube. In both (a) and (b) the outlet of the U tube is at the same level as that of the liquid in the reservoir. The pump uses the same amount of energy whether the U tube is horizontal or vertical. In the latter case the gravitational pressure of the liquid in the upward limb of the inverted U tube is counterbalanced by the gravitational pressure of the liquid in the downward limb (siphon principle). Thus the pump is not subjected to gravitational pressure of the column of the liquid. Note that in this case the total pressure of the curvature of the U tube will be equal to the dynamic pressure at the curvature + $(-\rho gh)$. It should be remembered that the height of the U tube is immaterial, except insofar as it increases the total resistance of the tube to flow. [Reproduced from Badeer, H.S.; Rietz, R.R.: Vascular hemodynamics: deeprooted misconceptions and misnomers. Cardiology *64*:197 (1979).]

ferent points in the vascular system due to changes in the *gravitational* component (not the dynamic) of blood pressure in the vessels. This occurs because the blood inside the vessels is freely subject to the action of gravity (weight of column of blood = height × density × acceleration of gravity), whereas the extravascular fluid (interstitial) is supported by dense connective tissue and the gel of interstitial fluid. Therefore, it is not free to exert an equivalent hydrostatic pressure on the outside of

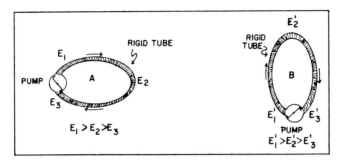

Fig. 159. Flow due to energy gradient created by pump. In a closed rigid tube system with circulating liquid, changes in the orientation of the system alter the transmural pressure at various points but that does not affect tube diameter; hence input, output, TPR, and total energy (pressure) gradients are unchanged. Therefore, energy expenditure of the pump is unaltered. Gravity neither helps nor hinders liquid flow in a closed circuit with rigid tubes.

the vessels. Changes in transmural pressure alter vascular diameters, hence their capacity and resistance to flow. When a person stands *erect* (orthostasis), his or her transmural pressure increases in *all* vessels (arteries, capillaries, veins) *below the level of the heart* (level of HIP). In the foot, the pressure in all vessels increases by about 88 mmHg (fig. 160). Veins distend most and hold extra (about 500 ml) blood, thereby temporarily decreasing the venous return from these vessels to the heart. In all vessels *above* the level of the heart (level of HIP), transmural pressure *decreases* and veins collapse, causing a temporary increase in venous return from these vessels (fig. 160). Note that the diameter of vessels inside the skull is not subject to significant change in response to postural changes, since the gravitational pressure in the blood columns is balanced by an equal change in the extravascular pressure of cerebrospinal fluid, surrounding these vessels (CSF is free to flow). The net change in *overall venous return* to the heart is an immediate decrease; hence the fall in cardiac output.

The fall in carotid sinus pressure from gravitational effects and from the *immediate* decrease in cardiac output brings into operation the carotid sinus and aortic reflexes, which increase TPR. This causes a decrease in the *vis a tergo*, which explains the *sustained decrease* in cardiac output in standing (fig. 161). In prolonged standing there is also a decrease in circulating blood volume caused by a loss of plasma into the interstitial spaces in the lower parts of the body.

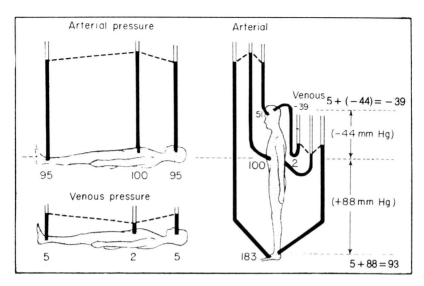

Fig. 160. The profound effect of posture on the mean arterial and venous transmural pressures (= dynamic + gravitational pressures). In the standing position, the numbers are estimated values of pressure, in mm Hg, exerted at those points with the transducer at the level where pressure is recorded. Note that the dynamic pressure drop (perfusion pressure) between the large arteries and large veins changes only to a slight extent as a result of changes in microvascular diameter (passive and active) and in blood flow. [Reproduced, with permission, from Burton, A.C.: Physiology and Biophysics of the Circulation; 2nd ed. (Year Book Medical Publishers, Chicago 1972).]

Fig. 161. Cardiovascular effects of assuming the upright position. Reduced venous return decreases central venous pressure, heart blood volume, pulmonary blood volume and cardiac output. Carotid sinus reflex causes peripheral vasoconstriction and reduces blood flow to skin, subcutaneous tissues, muscles, and splanchnic organs. Increased TPR restores aortic pressure to a more or less recumbent value. Diminished blood flow in areas mentioned causes increased oxygen extraction from the capillary blood. More blood (not blood flow) is in veins of the lower extremities. [Reproduced with permission from Rushmer, R.F.: Cardiovascular Dynamics; 2nd ed. (W.B. Saunders Co., Philadelphia 1961).]

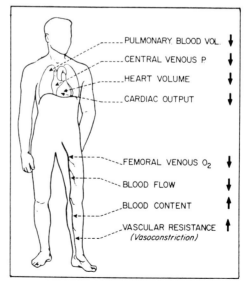

The carotid sinus and aortic reflexes serve to maintain the mean dynamic arterial pressure and cerebral blood flow.

Absence of Gravity in Space

In space, postural changes do not affect circulation.

The work of the heart is not affected by weightlessness except insofar as it causes changes in cardiac output and TPR. Blood has mass (although no weight) and pressure is created by the contraction of the heart to propel the mass of blood against the resistance of the peripheral vessels.

In the space capsule, there may be changes in cardiac output and TPR due to muscular inactivity, changes in blood volume, loss of fluid, changes in vasomotor tone, and so on.

References

Badeer, H.S.; Rietz, R.R.: Vascular hemodynamics: deep-rooted misconceptions and misnomers. Cardiology (Basel) *64:*197–207 (1979).

Burton, A.C.: Physiology and Biophysics of the Circulation; 2nd ed. pp. 97–100, 104–106 (Year Book Medical Publishers, Chicago 1972).

Caro, C.G.; Pedley, T.J.; Schroter, R.C.; Seed, W.A.: The Mechanics of the Circulation. (Oxford University Press, New York 1978).

Cardiovascular Regulation

22 Short-term Regulation

In everyday life the metabolic demands of tissues change constantly. These changes are related to acute changes in physical and mental activity of the subject, eg, muscular activity, food intake, changes in body posture, ambient temperature, and so on. Blood flow to various tissues must adjust *rapidly* to meet new demands. There are very efficient and complex mechanisms for achieving this. Such adjustments are described as short-term or *acute* regulation of circulation.

In contrast to this there are *long-term* regulations of the circulation which involve the regulation of body fluids and blood volume. In these, the kidney plays an important role, eg, hormonal control of water and salt excretion. Also the regulation of water and salt intake plays a significant role.

Short-term regulation may be considered under (a) peripheral mechanisms and (b) central mechanisms. Peripheral mechanisms consist chiefly of the metabolic control of blood vessels. Vasodilator metabolites and interstitial fluid P_{O_2} are the most important regulators (already discussed). The *central* or *reflex* mechanisms regulate the aortic pressure and the overall distribution of cardiac output by changing the vascular resistance in limited parts of the circuit (skin, skeletal muscle, splanchnic area, and so on). The relative importance of the peripheral and central regulation varies in different organs. In the most vital organs like the brain and the heart, metabolic regulation predominates which is a better safeguard to maintain nutritional blood flow. In less vital organs such as the skin and splanchnic organs, central regulation is more powerful. In skeletal muscle at rest central neural regulation predominates, whereas *during exercise* the metabolic control becomes more prominent.

Central or Reflex Regulation

Under this heading the following mechanisms will be briefly considered:

1. Carotid sinus reflex
2. Aortic arch reflex
3. CNS ischemic response

All of these are designed to maintain *brain circulation* by adjusting cerebral perfusion pressure

4. Acute reflexes arising from large veins and atria—to regulate heart rate and blood volume
5. Suprabulbar levels of cardiovascular regulation

Carotid Sinus Reflex

The carotid sinus is a swelling at the origin of the internal carotid artery characterized by being rich in elastic tissue and by the presence of special receptors in the tunica media and adventitia. These receptors are sensitive to stretch of vessel wall caused by the transmural pressure. The afferent nerve, called the carotid sinus nerve or Hering's nerve, joins the glossopharyngeal (IX) with its cell body in the *inferior* or *petrous* ganglion. The central fibers enter the medulla to make synaptic connections with the solitary tract nucleus which in turn sends fibers to the vasomotor and cardiac centers.

The simplest method to study the reflex is to isolate the sinus and fill it with static fluid at different pressures (fig. 162). Rise of sinus

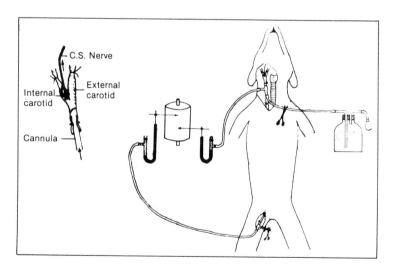

Fig. 162. Diagram illustrating carotid sinus perfusion. The internal and external carotid arteries are ligated and the common carotid is cannulated, leaving the carotid sinus nerve intact. The isolated blind sac is supplied with defibrinated blood and its internal pressure recorded with a mercury manometer. General arterial pressure is registered by a separate mercury manometer in the femoral artery.

pressure is followed by a fall in systemic arterial pressure and bradycardia, and vice versa (fig. 163). Note that the recovery of systemic pressure is not as rapid as its fall. Some refer to this as "unilateral rate sensitivity."

Study of the action potentials in the carotid sinus nerve has shown

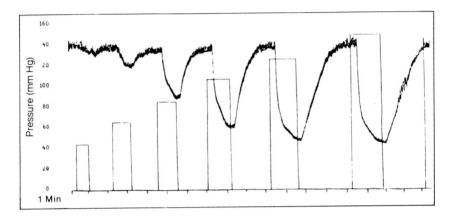

Fig. 163. The right carotid sinus in the rabbit is perfused by the method illustrated in the preceding diagram. The two vagi and aortic (depressor) nerves as well as the left carotid sinus nerve are cut. The rectangles represent the pressure in the right carotid sinus, where the nerve is intact. As this pressure is increased, the systemic arterial pressure falls correspondingly. [Reproduced, with permission, from Rein, H.; Schneider, M.: Physiologie des Menschen; 12th ed. (Springer-Verlag, Heidelberg 1956).]

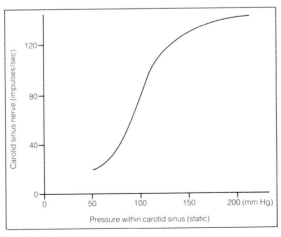

Fig. 164. Relationship of impulse frequency in the carotid sinus nerve to transmural pressure (nonpulsatile) across the wall of the carotid sinus. Note: threshold of firing is at about 50 mmHg and the maximum frequency is reached at about 200 mmHg.

that when the sinus pressure is below 40 or 60 mmHg, there are no impulses in the nerve. As pressure rises above this "threshold" value impulses in the nerve increase almost linearly, then the curve flattens out at about 200 mmHg. The most sensitive region of the curve is in the range of the normal mean blood pressure—80 to 120 mmHg (fig. 164). Different receptors have different thresholds. In a *single* fiber, the frequency of impulses increases with the rising pressure up to a maximum (fig. 165). In *acute* experiments the receptors show little "adaptation" to a constant stimulus (tonic receptors). This is useful in continuously monitoring blood pressure and maintaining it at a steady level.

However, if blood pressure increases *chronically* (hypertension) the receptors change their sensitivity. The curve shifts to the *right* within a

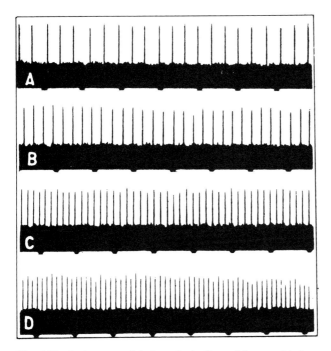

Fig. 165. Action potentials in a single fiber of the carotid sinus nerve of the rabbit. The carotid sinus was isolated and filled with blood to varying degrees—(A) sinus pressure held constant at 40 mmHg; (B) 80 mmHg; (C) 140 mmHg; (D) 200 mmHg. Time marks at 0.2 sec. Note the absence of rapid adaptation of the receptor to a constant stimulus (tonic receptor). Contrast this with the response of a touch receptor. [Reproduced, with permission, from Bronk, D.W.; Stella, G.: The response to steady pressure of single end organs in the isolated carotid sinus. Am. J. Physiol. *110:* 708 (1935).]

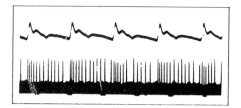

Fig. 166. Action potentials in the carotid sinus nerve under normal pulsatile changes of pressure in the carotid sinus. Upper record is arterial pressure. Time scale at the bottom is 1/5 sec. Note the increase in the frequency of impulses during systole. [Reproduced, with permission, from Bronk, D.W.; Stella, G.: Afferent impulses in the carotid sinus nerve. J. Cell Comp. Physiol. *1:*113 (1932).]

few days and the maximum sensitivity is at a higher blood pressure. In other words, the receptors are "reset" at a higher level. This is a useful "chronic" adaptation of receptors in that in a hypertensive patient the baroreceptors continue to function adequately in response to *acute* changes of sinus pressure, eg, changes in posture. Thus the baroreceptor system plays no role in the long-term regulation of arterial pressure.

Normally, the impulses in the carotid sinus nerve vary cyclically with the changes in arterial pressure with each heart beat (fig. 166). The receptors are more responsive to a rise in pressure than to a fall of pressure. In the medullary centers these impulses are "rectified" so that the *outflow* of impulses to blood vessels and the heart is *nonpulsatile*.

The impulses caused by stretch, on reaching the medulla (a) *inhibit* the vasoconstrictor center, which is stimulated to discharge by the normal arterial P_{CO2} and $[H^+]$ (pH = 7.4). The result is peripheral vasodilation (decrease in TPR), (b) they *stimulate* the cardiac vagal center and *inhibit* the cardiac sympathetic center, slowing the heart and reducing myocardial contractility. These tend to reduce cardiac output, which is not as important as the reduction in TPR in restoring blood pressure. The vascular and cardiac responses serve to reduce arterial pressure.

The reverse is also true, namely, any decrease or cessation of impulses in the carotid sinus nerve causes vasoconstriction, cardioacceleration and cardioaugmentation, raising blood pressure. If the fall in sinus pressure is rapid and marked, there is also the liberation of catecholamines from the adrenal medulla (eg, in acute severe hemorrhage).

The blood vessels involved in the carotid sinus reflex are those of the skin, skeletal muscles and the splanchnic area, including the renal vessels (brain and heart vessels being excluded).

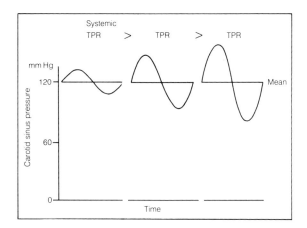

Fig. 167. Effect of the magnitude of pulse pressure in the carotid sinus on the tone of the vasoconstrictor center, keeping mean pressure constant. Increased pulse pressure increases the inhibition of vasomotor tone (causing a greater fall of arterial blood pressure).

An interesting observation is that if the *mean* pressure in the sinus is held constant but *pulse* pressure is altered, there is a greater reflex inhibition of vasomotor tone (vasodilation) at greater pulse pressures. In other words, at any given level of mean arterial pressure in the carotid sinus, the lower the *pulse pressure the greater the systemic vascular resistance* and vice versa (fig. 167). This characteristic tends to increase the effectiveness of the reflex in restoring blood pressure, eg, in hemorrhage blood pressure falls and pulse pressure is reduced; the reduced pulse pressure causes a greater reflex vasoconstriction than otherwise (useful to restore pressure).

The carotid sinus reflex is an excellent example of a negative feedback system (closed loop). There is a ''set point'' in restoring the aortic pressure to a constant level (fig. 168). What is the value of a constant aortic pressure? The answer is that it insures adequate blood flow to vital organs like the brain, spinal cord, and heart whenever something tends to disturb aortic pressure *acutely*. Less vital organs can temporarily alter their blood supply to insure the constancy of flow to the vital organs.

In everyday life the carotid sinus reflex plays an important role when one stands up from a recumbent position. Aortic pressure tends to fall from reduced venous return and cardiac output. Sinus pressure falls,

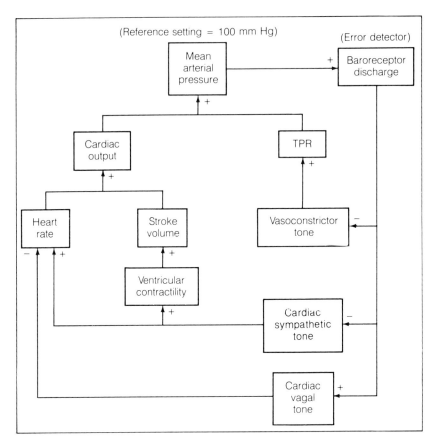

Fig. 168. Diagram of the negative feedback loop (closed) for the short-term regulation of arterial pressure by the baroreceptor reflexes (carotid sinus and aortic arch). The + sign and the − signs at the head of each arrow indicate the effect of augmenting the factor shown in the rectangle from which the arrow starts. + = increase; − = decrease.

also from gravity effects (weight of column of blood from heart to carotid sinus). The carotid sinus reflex immediately operates to maintain the aortic pressure and perfusion of the brain.

Physical inactivity for long periods causes the reflex to become sluggish, eg, prolonged stay in bed. Under these circumstances dizziness often occurs on standing up from the lying position. A similar problem is seen in astronauts returning to earth after a prolonged stay in orbit.

Clinical Applications
1. Treatment of paroxysmal tachycardia, which is a sudden marked increase in heart rate due to the rapid frequency of discharge of an ectopic pacemaker. *Strong* pressure on one of the carotid sinuses may inhibit the ectopic pacemaker by reflex vagal stimulation (A-Ch inhibits pacemaker activity!). This treatment often succeeds in atrial paroxysmal tachycardia but rarely in ventricular. The reason is that vagal fibers go to atrial muscle in large numbers but very few fibers reach ventricular muscle or Purkinje system. Remember that the *site* of pressure is very important. Pressure below the sinus on the common carotid artery is more than useless!
2. In tumors or arteriosclerosis of the carotid, the carotid sinus reflex may become hyperactive causing bradycardia, peripheral vasodilation and fainting when pressure is exerted on the sinus (tight collar, and so on). Treatment is to denervate the carotid sinus on the affected side.

Aortic Arch Reflex
Operates exactly like the carotid sinus reflex but is less powerful.

The aortic nerves were described in 1866 by de Cyon and Ludwig. The nerves are purely afferent, and in all animals except the rabbit, the nerves join the vagi (fig. 169). Stimulation of the central end of these nerves causes a fall in blood pressure, and bradycardia. The reflex operates like the carotid sinus but is less potent. Experimental studies are not as extensive because of the difficulties in perfusing the aortic arch.

The receptors in the carotid sinus and the aortic arch are called *pressoreceptors* or *baroreceptors*.

It is interesting to note that the acute regulation of arterial pressure by the baroreceptor buffer mechanisms (carotid sinus and aortic reflexes) is facilitated by certain hemodynamic alterations that occur. For instance, if arterial pressure rises as a result of an acute peripheral vasoconstriction (increased TPR) from any cause, this will be accompanied by a fall in end-capillary pressure (or *vis a tergo*), causing a drop in venous return and cardiac output. The drop in cardiac output will tend to counterbalance the effect of increased TPR on arterial pressure (Arterial Pressure = TPR↑ × CO↓), thereby aiding the restoration of arterial pressure. However, quantitatively the drop in cardiac output does not perfectly balance the influence of vasoconstriction (increased TPR)

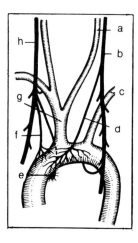

Fig. 169. Diagram showing the connection of the aortic nerves and body in the dog. a = Left common carotid artery; b = left vagus; c = left subclavian artery; d = left aortic nerve; e = aortic body; f = right aortic nerve; g = brachiocephalic artery; h = right vagus.

on arterial pressure. Thus, this hemodynamic mechanism is supplementary to the baroreceptor mechanisms in restoring arterial pressure.

Another complicating factor may also be considered. The initial vasoconstriction causes a drop in capillary pressure, which tends to shift interstitial fluid into the blood tending to increase blood volume and cardiac output. However, this mechanism is a comparatively slow process and does not play a significant role in the adjustment of arterial pressure under these circumstances.

If the rise in arterial pressure is due to an acute increase in blood volume and cardiac output (eg, intravenous injection of blood or saline), the baroreceptor buffer mechanisms will be supplemented by a shift from plasma into the interstitial fluid as a result of a rise in capillary pressure. This rise is due to reflex peripheral vasodilation as well as to the increase in blood volume. However, the fluid shift mechanism is a slow process, starting in about 5-10 minutes after injection and continuing for an hour or more. It is a negative feedback mechanism, because it tends to restore blood volume, venous return, cardiac output, and arterial pressure. Conversely, an acute decrease in blood volume (eg, hemorrhage) will cause fluid to shift from the interstitial compartment into the plasma as a result of the drop in capillary pressure (reflex vasoconstriction and decrease in blood volume).

CNS Ischemic Response or Cushing's Reflex

In 1901 Cushing noted that an acute rise in intracranial pressure by injecting fluid into the cerebrospinal (CSF) fluid causes a sharp rise in systemic arterial pressure. The hypertension is *maintained* at a constant level as long as the intracranial pressure is elevated and held constant (fig. 170). This means that the baroreceptor reflexes are unable to counteract and restore the blood pressure. In other words, the Cushing response is more potent than the *two* pressoreceptor reflexes working together.

The mechanism of Cushing's response seems to be through *ischemia* of the vasoconstrictor center in the medulla, which is stimulated to discharge intensely (probably the rise in P_{co_2} stimulates the vasomotor neurons).

The functional value of the CNS ischemic response is to safeguard the cerebral blood flow and brain function. Increased intracranial pressure compresses the cerebral vessels, which tends to reduce flow. The

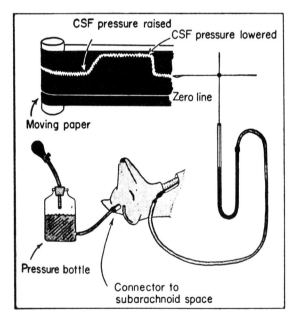

Fig. 170. Diagram illustrating the Cushing response. Rise of intracranial pressure (CSF pressure) causes a rise in arterial pressure which is maintained so long as the intracranial pressure is elevated. In other words, the carotid sinus and aortic reflexes together are unable to overcome the CNS ischemic response and restore the arterial pressure. [Reproduced, with permission, from Guyton, A.C.: Textbook of Medical Physiology; 5th ed. (W.B. Saunders Co., Philadelphia 1976).]

rise in arterial pressure elevates cerebral perfusion pressure and maintains flow:

$$\dot{Q} = \frac{P_1\uparrow - P_2}{R\uparrow}.$$

To be effective this mechanism must *override* the pressoreceptor reflexes, and *it does*.

Clinically this mechanism may function in acute intracranial hemorrhage from fractures of the skull. Two important signs are: (a) hypertension and (b) slow pulse (stimulation of vagal center). Any condition that increases intracranial pressure acutely causes these responses, but if the pressure develops slowly (eg, brain tumors), the hypertension is not so intense. Cerebral vascular diseases that cause ischemia of medulla also bring on hypertension (not common).

Reflexes from Large Veins and the Atria

For many years it has been postulated that an acute increase in venous inflow stretches the atria and reflexly causes cardioacceleration (Bainbridge reflex). Both afferent and efferent nerve fibers were thought to travel in the vagi. In recent years the existence of this reflex has been questioned.

Many investigators have shown that stretching the *isolated* or *denervated* S-A node increases its frequency of discharge (direct effect of stretch). Whether or not increased venous inflow *physiologically* can stimulate the S-A node directly is uncertain.

Acute increases in circulating blood volume increase the filling and pressure of the atria. There are stretch receptors, particularly at the junction of the pulmonary veins and left atrium, called volume receptors, which fire along afferents in the vagi, causing an increase in urine flow (diuresis). The atrial signals seem to reach the hypothalamus and reduce the secretion of antidiuretic hormone (ADH), which increases the excretion of water [Gauer and Henry 1963]. Also, atrial receptors seem to have reflex effects on kidney vessels (vasodilation of afferent arterioles?). The result is to reduce the secretion of *renin-angiotensin-aldosterone* causing more Na^+ and water excretion in the urine. This is a negative feedback mechanism to restore acute *increases* in blood volume. It should be noted, however, that in *chronic* hypervolemia these receptors are said to be "adapted" and are not very sensitive to increased blood volume.

Suprabulbar Levels of Cardiovascular Regulation

Experiments have shown that stimulation of many parts of the brain such as the cerebral cortex, thalamus, hypothalamus, and cerebellum produces changes in heart rate and blood pressure. In the cerebral cortex these areas include the frontal lobes, orbital cortex, motor and premotor areas, anterior part of temporal lobe and the limbic system (cingulate gyrus, hippocampus, amygdala). In the thalamus, stimulation of the midline, ventral, and medial groups of nuclei may cause tachycardia. In the hypothalamus, stimulation of the posterior part causes tachycardia and a rise in blood pressure, whereas stimulation of the anterior hypothalamus causes cutaneous vasodilation.

The exact role of these areas in various physiologic states in man is difficult, if not impossible, to determine.

It is well known that emotional stimuli have a profound effect on the heart and blood vessels. Strong psychic stress such as fear, anger, and excitement, produces tachycardia and a rise in blood pressure. Stimulation of the motor cortex and amygdala can evoke such cardiovascular responses. These reactions seem to be integrated at the posterior hypothalamic level causing excitation of sympathetic nerves to the heart and blood vessels. At the same time the sympathetic cholinergic vasodilator fibers to skeletal muscle are stimulated. These reactions serve to mobilize the organism for fight or flight (defense reactions). Under these circumstances cardiac work load is increased, and in patients with coronary artery disease sudden death may occur.

Some believe that frequent "low-grade" psychic stress in everyday life may play a role in the development of hypertension involving the hypothalamic sympathetic areas.

During muscular exercise there is evidence that the motor-premotor cortex activates pathways that induce cardioacceleration and vasoconstriction (eg, splanchnic area). Such changes seem to be relayed in the hypothalamus because lesions of the hypothalamus greatly reduce the tachycardia and the moderate rise in blood pressure during exercise.

According to some, assumption of the upright position activates the fastigial nucleus of the cerebellum, which sends excitatory impulses to the cardiac sympathetic and vasomotor centers in the medulla, inducing tachycardia and peripheral vasoconstriction.

The limbic-hypothalamic centers seem to be involved in the vaso-dilation of erectile tissue during sexual activity.

Other emotional reactions involve vasodilator mechanisms. One is *blushing,* which has not been satisfactorily explained. Another is *fainting* from psychic influences. In this condition there is vasodilation and bradycardia which probably involve the limbic system since stimulation of certain areas of the cingulate gyrus causes a fall in blood pressure, and bradycardia. The pathway passes by way of the anterior hypothalamus to act on the medullary centers, producing effects similar to stimulation of the carotid sinus and aortic nerves (inhibition of vasomotor tone and stimulation of vagal center). These cardiovascular effects are accompanied by reduced muscle tone and somatomotor activity with depressed breathing.

There are vasomotor changes in the skin (vasoconstriction) after the intake of food. It may explain the chilly sensation that occurs after meals, especially in the winter months. The neural mechanism is obscure.

Lastly, it may be noted that the chief centers for the regulation of body temperature reside in the hypothalamus. In this function the cutaneous blood vessels play an important role. Exposure to cold activates mostly the *posterior* hypothalamus, which induces cutaneous vasoconstriction and piloerection to conserve heat, and other reactions to increase heat production in the body (eg, shivering, and so forth). Exposure to heat activates chiefly the *anterior* hypothalamus, which induces cutaneous vasodilation, sweating, panting, and so on, to favor heat loss from the body. For details the reader must refer to chapters on temperature regulation in other texts.

References

Cushing, H.: Some experimental and clinical observations concerning states of increased intracranial tension. Am. J. Med. Sci. *124:*375–300 (1902).

Gauer, O.H.; Henry, J.P.: Circulatory basis of fluid volume control. Physiol. Rev. *43:*423–481 (1963).

Guyton, A.C.: Circulatory Physiology III: Arterial Pressure and Hypertension. (W.B. Saunders, Philadelphia 1980).

Heymans, C.; Neil, E.: Reflexogenic Areas of the Cardiovascular System. (J. & A. Churchill, Ltd., London 1958).

Kirchheim, H.R.: Systemic arterial baroreceptor reflexes. Physiol. Rev. *56:*100–176 (1976).

Special Circulations

23 Pulmonary Circulation

Pulmonary circulation is designed primarily for gas exchange between the blood and the alveolar gas rather than for the metabolic demands of lung parenchyma. Since metabolic regulation of blood flow is unnecessary, the vasomotor control is poor and the arterioles have thin walls with little or no smooth muscle (more elastic tissue). Consequently, the pulmonary arterioles are more distensible than the systemic. Critical opening pressure is lower than that of the systemic (<20 mmHg). A slight increase in right ventricular output and pulmonary arterial pressure, opens up many closed pulmonary arterioles resulting in a decrease in total pulmonary vascular resistance (eg, exercise).

The pulmonary circuit is a *low pressure* system due to the low resistance of its microvessels. Pulmonary artery pressure is about 25/10 mmHg. The true mean pressure is about 15 mmHg. Mean left atrial pressure is about 5 mmHg. Therefore, $P_1 - P_2 = 15 - 5 = 10$ mmHg. In the systemic circuit $P_1 - P_2$ is about 95 mmHg:

$$R = \frac{P_1 - P_2}{\dot{Q}}.$$

Since $\dot{Q}$ is the same for both circuits, then

$$\frac{R_{syst}}{R_{pulm}} = \frac{95}{10} = \frac{9 \text{ or } 10}{1}.$$

Thus, the TPR of the pulmonary vessels is about nine to ten times smaller than that of the systemic in a resting individual. Possible factors that may account for this would be (a) shorter length of arterioles, (b) larger radii of arterioles, and (c) greater number of parallel vessels.

The right ventricular wall is relatively thin and is not designed to pump against high pressures. The right ventricle is said to be a "volume" pump. Acute *diffuse* obstruction of the pulmonary microvessels (microembolism) causes an acute rise in pulmonary artery pressure, producing rapid failure of the right ventricle. *Chronic* gradual rise of pulmonary artery pressure produces right ventricular hypertrophy and ultimately failure (cor pulmonale, which refers to heart disease resulting from disease of the lungs).

Note that the *pulse* pressure in the pulmonary artery is about 15 mmHg as compared with the systemic of about 40 mmHg. This difference is due to the greater compliance of the pulmonary artery, compared with that of the aorta (fig. 171).

There is evidence (from body plethysmography) that in the intact resting subject, pulmonary capillary flow is *pulsatile,* which may be accounted for by the low resistance of the pulmonary arterioles.

Pulmonary capillary pressure can be estimated only indirectly. Since there is capillary flow during diastole, capillary pressure must be below pulmonary artery diastolic pressure (about 10 mmHg). Also, it must be higher than left atrial pressure, which is about 5 mmHg. The approximate normal value is taken to be around 8 mmHg. The pulmonary interstitial fluid pressure is generally believed to be close to the mean alveolar pressure, which is about zero. The oncotic pressure of plasma is about 25 mmHg. Therefore, there is no filtering force; on the contrary, there is a force to absorb interstitial fluid ($25 - 8 = 17$ mmHg). This tendency to absorb is referred to as the "factor of safety" to pre-

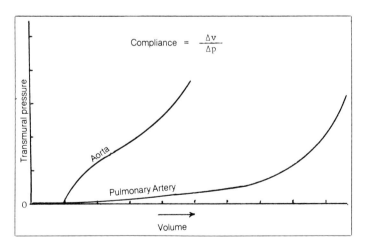

Fig. 171. Diagram comparing the *in vitro* compliance of the pulmonary artery with that of the aorta.

vent lung edema. If the alveoli are filled with a saline solution, there is rapid absorption of the fluid. If water enters the alveoli (eg, fresh water drowning), a very rapid absorption occurs, which may cause osmotic hemolysis of red blood cells.

If we accept the thesis that the pulmonary capillaries do not filter, what would be the origin of the pulmonary interstitial fluid and lymph? This may be derived from the *bronchial* circulation which presumably has a high capillary pressure since it belongs to the systemic circuit derived from the aorta. Pulmonary lymph has a high protein content (about 3%–4%). The origin of this protein is not clear; it may be derived from the bronchial capillaries (?). Lymph vessels of the lungs do not extend beyond the level of the alveolar ducts. Note that the bronchial capillaries drain into the pulmonary venules and veins.

According to Guyton [1981], the pressure of interstitial fluid surrounding the pulmonary capillaries is about −6 mmHg and its protein is derived from pulmonary capillaries. This protein has an oncotic pressure of about 16 mmHg. Using these values for capillary pressure and oncotic pressure of plasma, one would get a net outward force of 14 mmHg [8−(−6)] and a net inward force of 9 mmHg (16−25). This would imply continuous filtration of fluid across the pulmonary capillaries rich in proteins and is believed to be carried away by the lymphatics as rapidly as it is formed. The difficulty with this concept is that there would be very little "factor of safety" for edema formation in the lungs. Clinical and experimental observations do not support such a concept, eg, left ventricular failure does not readily cause pulmonary edema unless the capillary pressure rises excessively.

Arteriovenous anastomoses have been described for the pulmonary vascular bed (500 μm diameter glass microspheres can pass through the isolated human lungs). Their functional significance is not clear. If these vessels do not exchange gases and are open to circulation, they would cause shunting of unsaturated blood and reduce arterial blood oxygen saturation (venous admixture).

On assuming the upright position, the pressure in the apical lung vessels falls and that of basal vessels increases (due to gravity). As a result, the apical vessels collapse increasing their resistance thereby causing a decrease in blood flow. The reverse occurs at the base of the lungs (figs. 172 and 173). The ratio of alveolar ventilation to blood perfusion ($\dot{V}_A$)/($\dot{Q}$) which is normally around 0.8 becomes greater than 1.0 in the apex because of the marked decrease in blood flow ($\dot{Q}$).

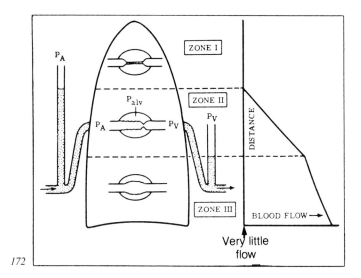

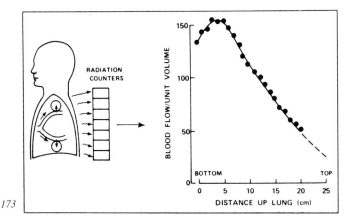

Fig. 172. Effect of gravitational pressure of blood on the diameter and resistance of pulmonary blood vessels and the regional differences in blood flow in the erect posture. Very little blood flows through the apex, increasing progressively towards the base. [Reproduced, with permission, from Selkurt, E.E.: Physiology; 4th ed. (Little, Brown and Co., Boston 1976).]

Fig. 173. Measurement of the distribution of blood flow in the upright human lung using radioactive xenon injected into the right heart. The dissolved xenon is evolved into the alveolar gas from the pulmonary capillaries. The units of blood flow are such that if flow were uniform, all values would be 100. Note the small flow at the apex. [Reproduced, with permission, from West, J.B.: Respiratory Physiology—The Essentials; 2nd ed. (Williams and Wilkins Co., Baltimore 1979).]

During muscular exercise in man, there is a slight rise in the mean pulmonary arterial pressure which is much less than the increase in cardiac output (fig. 174). If the output increases 200% (5 → 15 liters/min) mean pulmonary arterial pressure may rise only 50% (15→22 mmHg). This means that there is a *decrease* in pulmonary TPR. The decrease is believed to be due to dilation of open vessels and opening up of closed ones (recruitment), specially the apical. It seems to be a *passive* dilation as a result of the rise in pulmonary artery pressure and fall of extravascular pressure from the greater expansion of the lungs during exercise (greater negativity in the pleural space). The normal flow gradient from apex to base in the upright position tends to be reduced. If a lung is removed (pneumonectomy), there is very little rise in pulmonary pressure because the closed vessels of the remaining lung open up readily, reducing their resistance to flow.

The pulmonary blood volume (not flow rate) has been estimated to be about 10% of the total circulating blood volume (c̄ā 500 ml) in the upright position. Of this, about 80–100 ml is believed to be in the pulmonary capillaries. The rest is equally divided between the pulmonary arteries (200 ml) and the pulmonary veins (200 ml).

Pulmonary blood volume is increased acutely during muscular exercise and when one lies down (+ 200 ml). It increases chronically in

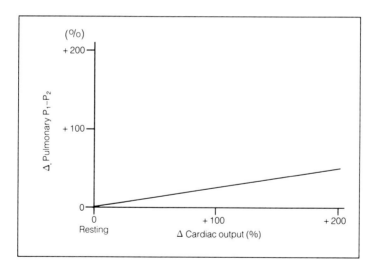

Fig. 174. Changes in cardiac output and pulmonary arterial pressure during muscular exercise.

mitral stenosis and in chronic left ventricular failure. In these disease conditions, the pressure throughout the entire pulmonary circuit is increased to overcome the *vis a fronte* of the left heart. This causes *passive* congestion in the lungs with an increase in pulmonary blood volume. In such patients, exercise or lying down causes additional pulmonary congestion and, if severe, may cause dyspnea or shortness of breath (early symptom of left ventricular failure).

High levels of pulmonary artery pressure tend to increase the circumference of the pulmonary artery valve ring because the artery is thin walled. This results in a "functional" pulmonic regurgitation (causing a diastolic murmur in the pulmonic area known as Graham Steell murmur).

It is interesting to note that *chronic* pulmonary hypertension causes thickening of the lung vessels as an "adaptive" response. Whether or not this is reversible is uncertain.

Control of Lung Vessels

May be discussed under neural, chemical and mechanical influences.

Neural. Lung tissue has a low metabolic rate and its nutrition is mostly derived from the bronchial circulation. Hence, pulmonary circulation is not designed to adjust flow according to metabolic needs of the lungs. Vasomotor control is rather weak. Stimulation of sympathetic nerves and norepinephrine cause a slight pulmonary constriction (α receptor) and a slight rise in pulmonary arterial pressure.

Vagal stimulation or injection of ACh into the pulmonary artery causes a slight decrease in pulmonary vascular resistance and a fall in pulmonary artery pressure.

Chemical. In contrast to systemic vessels, metabolites derived from lung tissue play no role in altering vascular resistance to adjust pulmonary blood flow. On the other hand, alveolar P_{O_2} has an action on pulmonary microvessels opposite to that on the systemic.

Low alveolar P_{O_2} has a *local arteriolar constrictor* action (contrary to its action on systemic arterioles). The significance of this seems to be to match alveolar blood perfusion to alveolar ventilation, eg, poor alveolar ventilation in localized areas of the lungs reduces blood flow to

SEA–LEVEL   HIGH ALTITUDE

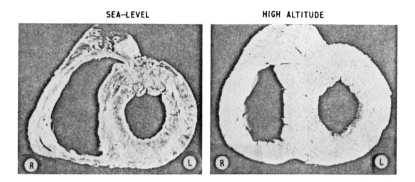

Fig. 175. Hypertrophy of the right ventricle at high altitudes. Cross section of the ventricles in a subject residing at high altitude (21-year-old man) as compared with that of a resident at sea level (11-year-old boy). Note the marked hypertrophy in the subject from high altitude. [Reproduced, with permission, from Recavarren, S.; Arias-Stella, J.: Right ventricular hypertrophy in people born and living at high altitudes. Br. Heart J. 26:806 (1964).]

these alveoli and shunts the blood to well-ventilated areas. This reaction tends to minimize the venous admixture of the arterial blood (reduces arterial unsaturation).

However, sometimes this response creates problems—eg, if alveolar hypoxia is *diffuse* throughout both lungs—as in residents of very high altitudes, the pulmonary vasoconstriction may have detrimental effects. Plumonary hypertension and right ventricular hypertrophy are common in man and animals living at high altitudes. Cattle are particularly sensitive and may develop hypertrophy and failure of the right ventricle (Brisket disease). Fortunately, it is rare in man (fig. 175). Part of the increased pulmonary vascular resistance may be due to the polycythemia (hematocrit = 60%). However, systemic blood pressure is normal or below normal because peripheral microvessels *dilate* and *open up* from arterial hypoxemia (increased capillarization ?) offsetting the effect of increased blood viscosity. In *long-term* residents of high altitude, heart rate, cardiac output and systemic TPR are not significantly altered. Therefore, the left ventricle is often not hypertrophied.

Mechanical. Pulmonary vessels do not exhibit the myogenic response to stretch or increased tension.

The lack of metabolic control and the lack of myogenic response explain why the pulmonary circulation does not demonstrate the phenomenon of autoregulation of blood flow. What would be the disadvantage if it did? Consider the situation in muscular exercise.

References

Adams, W.R.; Veith, I.: Pulmonary Circulation (Grune and Stratton, New York 1959).

Cumming, G.: The pulmonary circulation; in Guyton and Jones: Cardiovascular Physiology; MTP International Review of Science; pp. 93–122 (University Park Press, Baltimore 1974).

24 Fetal Circulation and Circulatory Changes at Birth

In the fetus, the right ventricle receives its blood mostly from the head and upper extremities through the superior vena cava, but some blood comes from the coronary sinus and a small amount from the inferior vena cava. The output of the right ventricle divides into *two:* about one-third goes to the lungs and two-thirds runs through the *ductus arteriosus* into the aorta because the aortic pressure is about 5 mmHg less than that of the pulmonary artery (70 PA vs. 65 mmHg aortic) and the resistance of the ductus is low. For this reason, in the fetus, the pulmonary and the systemic circuits are said to be *in parallel* (fig. 176). The relatively small flow through the lungs is due to the high resistance of the pulmonary microvessels. This may be related to the poor O_2 content (52% saturation) of the blood in the pulmonary artery, and hypoxia is known to constrict the lung microvessels. The continuous constriction may account for the thick muscular arterioles of the fetal lung (?). It is also likely that the presence of fluid in the fetal lungs compresses the microvessels and increases their resistance to blood flow.

The inferior vena cava receives blood from the lower extremities, from the portal system through the hepatic veins, and from the *single umbilical vein* by way of the ductus venosus (umbilical vein also gives off branches to the liver, which drain into the inferior vena cava by way of the hepatic veins). Blood from the inferior vena cava is directed chiefly into the left atrium through the foramen ovale. There, it mixes with blood from the pulmonary veins, enters the left ventricle and is pumped into the aorta. About one-third of the blood flowing up the ascending aorta goes to the head and upper extremities and the remainder goes down to the rest of the body after mixing with blood from the ductus arteriosus. A major fraction of the blood flowing in the descending aorta goes to the placenta by way of the *two umbilical arteries,* which arise from internal iliacs.

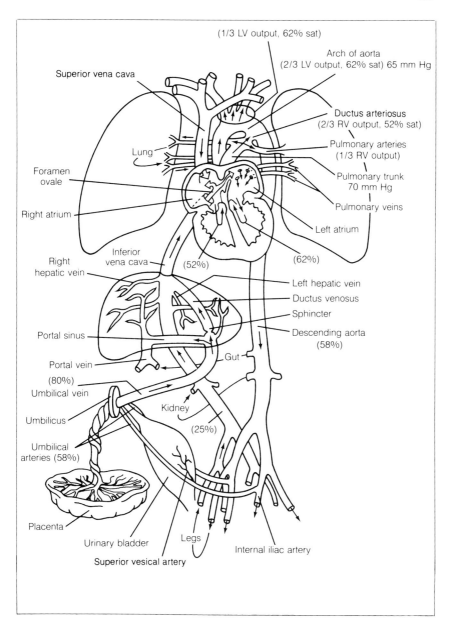

Fig. 176. Schematic diagram of the fetal circulation. The percentages are the oxygen saturation of the blood. Organs are not drawn to scale. [Modified from Moore, K.L.: Before We Are Born: Basic Embryology and Birth Defects; 2nd ed. (W.B. Saunders Co., Philadelphia 1983).]

Blood in the umbilical arteries is about 58% saturated with oxygen and becomes about 80% saturated in the umbilical vein after taking up oxygen in the placenta. When blood from the umbilical vein mixes with portal blood and with blood from the lower extremities (25% saturated) and lungs and reaches the left ventricle, its oxygen saturation goes down to about 62%. Blood from the right ventricle is about 52% saturated. Mixing of the ejected blood of the left ventricle (62% saturated) with that of the right ventricle (52% saturated) in the descending aorta gives rise to a saturation of about 58% as mentioned above (slightly more blood comes from the left ventricle). This is the saturation of blood going to the placenta, which raises the saturation to 80%.

At birth, when flow through the umbilical vessels is arrested, a number of changes occur:

1. The ductus venosus closes (mechanism—fall of pressure?).
2. Breathing begins due to asphyxia and cooling of skin (perhaps the obstetrician's slap).
3. Reduction of pulmonary vascular resistance to about one-tenth its previous value. The mechanism seems to be the filling of the lungs with gas instead of fluid. Apparently, it is not due to oxygen in the alveolar air because the same drop is said to occur if the lungs are inflated with nitrogen. On the other hand, filling the lungs with liquid prevents the reduction of pulmonary vascular resistance. The pulmonary artery pressure drops to about 35 mmHg and blood flow through the lungs increases, despite the fall in pulmonary perfusion pressure.
4. Left atrial pressure rises due to increased flow from the lungs, serving to close the foramen ovale.
5. The pressure gradient and flow between the pulmonary artery and the aorta *reverses* (this is chiefly due to the fall of pulmonary artery pressure with only a slight rise in aortic pressure due to occlusion of the umbilical arteries).
6. Constriction of the ductus appears to be initiated by the high P_{O_2} of arterial blood passing through it (52% becomes 95% saturated). Also, bradykinin formed from kininogen in the lungs when they fill with air may contribute to the contraction of the ductus. Narrowing of the ductus produces a turbulent flow (high velocity), which frequently causes murmurs in the newborn. Permanent closure of the ductus may take several weeks.

At birth the two ventricles have about the same wall thickness. With the fall in pressure in the pulmonary artery, the right ventricle gets thinner and the smooth muscle of the pulmonary arterioles decreases over a period of months [less active smooth muscle due to high alveolar P_{O_2} (?)]. In contrast, the left ventricular wall gets thicker due to the rise in systemic arterial pressure with age and body growth, until the ratio becomes 3:1 in the adult.

References

Berne, R.M.; Levy, M.N.: Cardiovascular Physiology; 4th ed.; pp. 248–250 (C.V. Mosby Co., St. Louis 1981).

Dawes, G.S.: Foetal and Neonatal Physiology (Year Book Medical Publishers, Chicago 1968).

Walsh, S.Z.; Meyer, W.W.; Lind, J.: The Human Fetal and Neonatal Circulation: Function and Structure (C.C. Thomas, Springfield, Illinois 1974).

25 Coronary Circulation

The large extramural coronary arteries run on the surface of the heart and then dip into the myocardium at right angles, branching profusely as the intramural vessels (fig. 177). The veins accompany the arteries. Venous blood empties into the right atrium by way of the coronary sinus and other smaller veins. Some very small veins empty directly into *all four* chambers of the heart, referred to as the Thebesian vessels. If the number of muscle fibers and the number of injected capillaries in cross section are counted, the ratio is about 1:1.

Normally, there are intercoronary anastomoses of arterioles with diameters between 20 and 350 μm, but no anastomoses in larger arteries. Therefore, sudden *complete* occlusion of a *normal* large coronary artery leads to death (necrosis) of the myocardium supplied (infarction). With light microscopy, detectable changes in muscle occur only after 4 to 8 hours of occlusion. On the other hand, if narrowing of the arteries occurs very *gradually* (as in disease) and then complete occlusion supervenes, there may be no infarction. This is because the anastomotic vessels have enlarged prior to occlusion and thereby maintain an adequate blood supply to the muscle after the main artery is occluded. The *stimulus* to the enlargement of the anastomotic vessels is not well understood. One concept is that the metabolites from the partially ischemic tissue and/or decreased P_{O_2} cause sustained vasodilation which increases wall tension in the anastomotic vessels inducing hypertrophy of the vessel wall. Another view is that pressure falls distal to the narrowing. This causes an increase in flow through the anastomotic channels, which somehow leads to vascular enlargement.

Measurement of Coronary Flow

The three principal methods that have been used are: (a) nitrous oxide, (b) indicator washout, and (c) velocity transducer at the tip of a catheter. The details of these methods are described in Appendix 3.

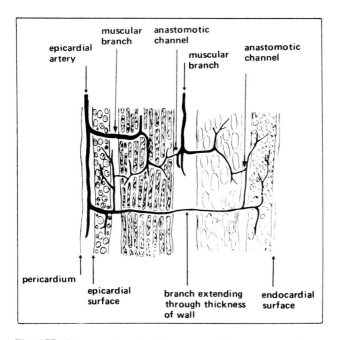

Fig. 177. Diagram showing the course of the coronary arteries as they run on the surface of the heart and penetrate at right angles into the myocardium, giving off branches between the various layers of muscle. [Reproduced, with permission, from Katz, A.M.: Physiology of the Heart (Raven Press, New York 1977).]

With the N_2O method in man at rest, the coronary flow is about 70–80 ml/100 g per minute (average heart weight is about 300 g). Thus, the heart in a resting individual receives about 5% of the cardiac output (250/5000). During muscular exercise coronary flow may become four or five times the resting value.

Factors Determining the Coronary Flow

Flow varies directly with the perfusion pressure and inversely with the total vascular resistance:

$$\dot{Q} = \frac{P_1 - P_2}{R}.$$

Perfusion Pressure

P_1 is the aortic pressure at the orifices of the coronary arteries and P_2 is the right atrial pressure. Both of these pressures are continuously fluctuating with the cardiac cycle. To alter P_1 experimentally without influencing cardiac load and metabolic rate, requires perfusion of the coronary arteries with blood from a reservoir. Figure 178 shows the results of such a study. The critical closing or opening pressure in the beating heart is about 20 mmHg. There is some degree of autoregulation of blood flow between 50 and 130 mmHg. The metabolic hypothesis is probably the explanation!

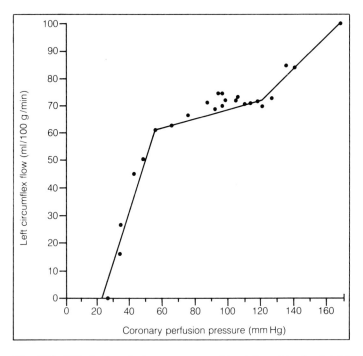

Fig. 178. Effect of perfusion pressure on blood flow (steady state) in a coronary artery when cardiac output, aortic pressure and heart rate are kept constant. Note the critical opening pressure and the autoregulation of blood flow within a limited range of perfusion pressure. The artery was perfused by a separate system, independent of aortic flow. [Reproduced, with permission of the American Heart Association, from Mosher, P.; Ross, J., Jr.; McFate, P.A.; Shaw, R.F.: Control of coronary blood flow by an autoregulatory mechanism. Circ. Res. *14:*250 (1964).]

Resistance of Coronary Vessels

Recall that total vascular resistance depends on the length, diameter and total number of functioning small vessels and the viscosity of blood. In the coronary vascular bed the number of "open" capillaries is controversial. The classic view is that all capillaries are open under resting conditions, but recent work suggests this may not be true.

Coronary vascular resistance is controlled by a number of influences that may be grouped under (a) nervous, (b) chemical, and (c) mechanical.

Nervous Control. Coronary vessels receive fibers from the vagi and from the cardiac sympathetics. The *direct* action of these nerves is not powerful and is somewhat controversial.

Sympathetic nerves seem to constrict the large coronary vessels (α-receptors) and dilate the small vessels (β-receptors). These studies were conducted in an arrested or fibrillating heart with perfused coronary arteries to prevent changes in flow secondary to metabolic alterations of the myocardium (eg, changes in heart rate, contractility, and so on).

The vagus nerves (cholinergic) seem to have a weak, direct dilator action on the coronary vessels, but this is not firmly established.

In the *beating* or pumping heart, the direct effect of stimulating these nerves is masked by the *more powerful* indirect effects produced by metabolic changes secondary to changes in heart rate, force of contraction, and so on. Thus cardiac sympathetic nerve stimulation with constant aortic pressure causes a marked *increase* in coronary blood flow (cardioacceleration, positive inotropic effect produces an increased metabolic rate and dilator metabolites). Vagal stimulation with constant arterial pressure tends to *reduce* coronary flow, probably because of the overriding influence of reduced myocardial metabolism from bradycardia, slight negative inotropic effect, and so forth.

Chemical Influences. It is generally held that the metabolic products from the contracting myocardium constitute the most potent regulators of coronary vascular resistance. *Coronary blood flow is more or less adjusted to the metabolic rate of the myocardium.*

In the heart-lung preparation, administering 5%–10% CO_2 in the inspired air causes a moderate increase in coronary flow (heart rate and blood pressure being held constant). Likewise, injection of lactic acid into the blood increases coronary flow moderately. Also, histamine, ATP and adenosine are coronary dilators. None of these has been proven to be *the* coronary vasodilator metabolite.

Decreased arterial P_{O_2} (normal $= 95-100$ torr) reduces coronary vascular resistance markedly. According to some, reduced P_{O_2} relaxes vascular smooth muscle by a *direct* action. Others claim it is *indirect* through the accumulation of some metabolite from heart muscle. Berne and Rubio [1979] suggest it is *adenosine*, but this is not accepted by all investigators. Whatever the mechanism, increased coronary flow serves to maintain the O_2 supply to the myocardium. Thus, *moderate* arterial hypoxemia does not cause a decline in the O_2 uptake of the heart because the increased coronary flow provides adequate amounts of oxygen to the myocardium. However, extreme degrees of arterial O_2 desaturation reduce cardiac O_2 consumption and cause myocardial failure.

Reactive hyperemia is very marked in the myocardium due to the high metabolic rate of the pumping heart. The normal heart uses 8 to 10 ml oxygen/100 g per minute.

Mechanical Factors. During systole the contracting myocardium of the left ventricle compresses the intramural vessels and impedes arterial blood flow (a kind of throttling effect) (fig. 179). Simultaneously, the outflow from the coronary sinus and other veins is increased. The reverse occurs during diastole. The throttling effect of systole is very little in the myocardium of the right ventricle due to the low pressure in that chamber (fig. 179).

The overall effect of normal rhythmic contractions of the heart on the coronary *flow per minute* is to *reduce it*. Gregg [1963] showed this by perfusing the coronaries and stopping the heart by vagal stimulation or injecting K^+ or by inducing ventricular fibrillation ($P_1 - P_2$ being kept constant) (figs. 180 and 181). Apparently, in the pumping heart the mechanical compression during systole is more potent in reducing flow than is the increase in vasodilator metabolites derived from contractile activity.

Gregg studied coronary flow in the large epicardial arteries with the electromagnetic flowmeter in unanesthetized dogs (fig. 182). Under resting conditions, about 25% of the total flow/cycle occurs during systole and 75% during diastole. This is partly due to the duration of each phase. *Moderate* cardioacceleration tends to increase coronary flow despite the increased frequency of extravascular compression. Apparently, here the dilator action of increased metabolites overrides the effect of

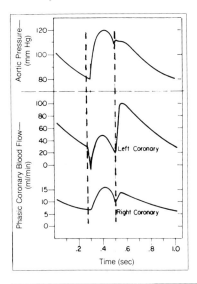

Fig. 179. Changes in coronary blood flow in the right and the left arteries during the cardiac cycle. Note the throttling effect of systole on the left coronary flow as a result of the high intramyocardial pressure, whereas the right coronary flow is affected little (low right ventricular pressure, hence little extravascular compression during systole). [Modified from Berne, R.M.; Levy, M.N.: Cardiovascular Physiology; 4th ed. (C.V. Mosby, St. Louis, Mo. 1981).]

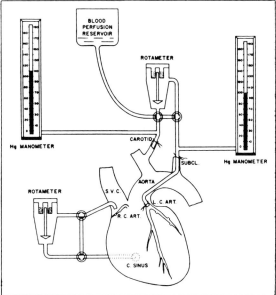

Fig. 180. Perfusion of a coronary artery. A metal cannula is placed in the left coronary artery and a catheter in the coronary sinus. Both are connected to flowmeters (rotameters) for recording arterial inflow and venous outflow. The left coronary artery is perfused with oxygenated blood from a reservoir placed at an adequate height (pressure). [Reproduced, with permission of the American Heart Association, from Sabiston, D.C., Jr.; Gregg, D.E.: Effect of cardiac contraction on coronary blood flow. Circulation *15:* 14 (1957).]

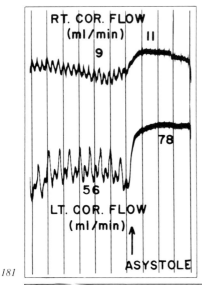

181

Fig. 181. Both the right and the left coronary arteries are perfused by gravity in a pumping dog heart. At arrow, the heart was stopped by ventricular fibrillation. Note the effect of asystole in increasing coronary arterial flow (removal of extravascular compression). Time lines = 1 sec. [Reproduced, with permission, of the American Heart Association from Sabiston, D.C., Jr.; Gregg, D.E.: Effect of cardiac contraction on coronary blood flow. Circulation *15:*14 (1957).]

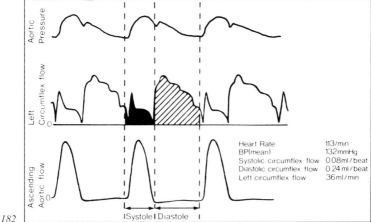

182

Fig. 182. Unanesthetized dog standing at rest (28 days post-op). Flow measured by electromagnetic flow meters. [Redrawn from Gregg, D.E.: Coronary blood supply and oxygen usage of the myocardium.; in Dickens and Neil: Oxygen in the Animal Organism (Pergamon Press, New York 1964).]

compression and shortening of the period of diastole. If heart rate is very fast, say 200/min, there may be inadequate coronary flow in relation to the metabolic demand of the heart. This may cause heart failure, if such tachycardia lasts for days or weeks.

Recent interest has centered on the regional distribution of coronary flow to the subendocardial compared with the subepicardial layers of the left ventricular myocardium. Pressure is much higher in the subendocardial layers of the left ventricle *during systole,* which tends to diminish the flow. Studies of flow/unit mass of tissue going to the deep, compared with the superficial, layers of the left ventricle in dogs have given conflicting results. Injection of radioactive microspheres (9μm in diameter) into the coronary artery and study of its distribution in the deep and superficial layers of the left ventricle show *equal distribution.* This suggests that normally any interference with flow to the subendocardial layers during systole is compensated by a greater flow during diastole, so that overall *flow per cycle* is approximately equal in the various layers. However, there is some evidence to suggest that the O_2 requirement of the subendocardium is greater than that of the subepicardium due to the greater contractile tension of the subendocardium resulting from the greater transmural pressure ($T = P \cdot r$). This view is supported by the lower P_{O_2} of the tissue fluid and of the venous blood from the subendocardium, and also by the greater concentration of lactate or of the lactate/pyruvate ratio.

In coronary artery disease or other pathologic conditions (eg, cardiac hypertrophy), the endocardial layers may receive less blood/unit mass of tissue than the epicardial under resting condition thus becoming more prone to ischemic damage (infarction). Such a discrepancy may become exaggerated during muscular activity.

A characteristic feature of the coronary circulation is the very low O_2 content (or saturation) of coronary venous blood. In man, it is about 5 vol%. If the arterial blood O_2 content is 19 vol%, the a-v O_2 difference will be 14 vol%. This is the highest extraction of oxygen among all organs in the body under resting conditions. It is probably due to the pumping activity of the heart which: (a) reduces overall coronary flow as compared with the nonbeating heart, and (b) increases the O_2 demand of heart muscle.

Both of these lead to greater extraction. A similar situation occurs in skeletal muscle. Resting skeletal muscle has a venous O_2 of about 13 vol%. During strong rhythmic activity, venous O_2 saturation from active muscles falls to about 5 vol%.

During muscular exercise, the coronary venous oxygen content *does not* fall appreciably, or falls very little. Hence, the increased oxygen demand of the heart is met by a proportional increase in coronary blood

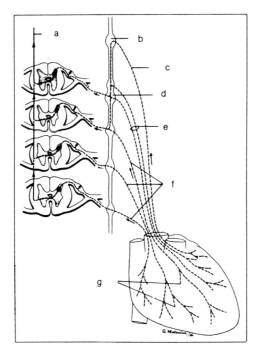

Fig. 183. Course of sensory fibers from the myocardium mediating cardiac (anginal) pain. a = lateral spinothalamic tract; b = middle cervical ganglion; c = middle cardiac nerve; d = stellate ganglion; e = inferior cardiac nerve; f = direct thoracic cardiac nerves; g = metabolites of hypoxia appear to constitute the adequate stimulus.

flow. This is true not only in exercise but also in other physiologic circumstances that increase myocardial oxygen demand, eg, rise of blood pressure due to emotional factors, severe cold exposure, and so forth. Consequently, *coronary blood flow normally varies in almost direct proportion to the myocardial O_2 demand*. If the coronary vessels are thickened and narrowed markedly, coronary flow may not increase adequately when the myocardial demand is sharply increased, and a serious discrepancy between the two may arise—eg, exercise, exposure to severe cold, after a heavy meal, emotional upsets, and so on. Under these conditions an ischemic type of pain may develop in the chest called angina pectoris. The afferent pathways are in the cardiac sympathetics and ascend to the thalamic centers concerned with visceral pain (fig. 183). The pain is an important early warning of coronary artery disease but promptly disappears when all activity is stopped and a rapidly acting vasodilator is taken.

References

Berne, R.M.; Rubio, R.: Coronary circulation; in Berne, Sperelakis and Geiger: Handbook of Physiology; sect. 2: The Cardiovascular System; vol. 1; The Heart; pp. 873–952 (American Physiological Society, Washington, D.C. 1979).

Gregg, D.E.; Fisher, L.C.: Blood supply to the heart; in Hamilton and Dow: Handbook of Physiology; sect. 2; Circulation; vol. II; pp. 1517–1584 (American Physiological Society, Washington, D.C. 1963).

Griggs, D.M.: Blood flow and metabolism in different layers of the left ventricle. The Physiologist *22:*36–40 (1979).

Marcus, M.L.: The Coronary Circulation in Health and Disease (McGraw-Hill Book Co., New York 1983).

Mohrman, D.E.; Feigl, E.O.: Competition between sympathetic vasoconstriction and metabolic vasodilation in the canine coronary circulation. Circ. Res. *42:*79–86 (1978).

26 Cerebral Circulation

Anatomic Features

The arterial blood supply of the brain is from the internal carotid and the vertebral arteries which anastomose at the base of the brain to form the *circle of Willis*. From this anastomosis arteries arise to supply the cerebral hemispheres, the mid-brain, the pons, the cerebellum and the medulla. Caudally, the vertebrals give off the anterior and posterior spinal arteries and continue upward to unite and form the basilar artery. All of these arteries penetrate into the nervous tissue from the external surface and carry the *pia* mater to variable distances.

The size and functional significance of the *internal carotid* artery vary with the *degree of encephalization* in a given animal species. In man, the carotid is the major source of blood to the brain. However, because of anastomosis, complete occlusion of *one* internal carotid in a person with *normal* arteries has no detectable effect on brain function. Complete occlusion of *both* internal carotids may lead to functional disturbances, especially if the arteries are diseased. (Clinical rule: never attempt to press the carotid sinuses *bilaterally* to treat paroxysmal tachycardia.) The arteries on each side supply the corresponding half of the brain with little mixing of blood between the two sides, because pressures are about equal in the corresponding vessels of the circle of Willis.

A unique feature of the circulation of the brain and of the spinal cord is that the vessels and other contents are enclosed in a *rigid* container. Since the contents are incompressible, any change in the volume of any one component must be accompanied by an opposite change in another. Most likely to change is the volume of *blood* in the vessels and of the *cerebrospinal fluid* (CSF). The relation between the *total* volume of the contents and the capacity of the cranium and the vertebral canal creates the normal *intracranial* or *spinal pressure* exerted on all structures within the system (it is a kind of tissue pressure). In the horizontal position of the body, this pressure is about 10–15 cm H_2O. When a person stands erect, pressure in the cranial cavity drops to below atmospheric (about -20 cm H_2O), while that at the lower vertebral canal rises to about $+50$ cm H_2O. The pressure is about zero (or atmospheric) at the level of the neck or cysterna magna. These changes are due to the action of gravity on the column of CSF and the fact that the container is rigid. These changes in CSF pressure with posture eliminate most of the effects of postural changes on the diameter and resistance of the *vessels of the brain and spinal cord*, because intravascular and extravascular pressures change almost equally. The change in the pressure of the column of blood inside these vessels is balanced by a similar change in the pressure of CSF on the outside of the vessels. Hence, in vivo *transmural* pressure and vascular resistance are practically unchanged. Likewise, accelerating forces do not disturb cerebral and spinal vascular diameters and flow resistance as much as they do those of other vessels in the body.

Measurement of Cerebral Blood Flow

1. N_2O method—same as that of coronary (see Appendix 3).
2. Indicator washout with isotopes (^{85}Krypton or ^{133}Xenon) (see Appendix 3).
3. Regional blood flow has been estimated by autoradiography in animals or by γ-scintillation monitoring systems in man with ^{133}Xe (Ingvar 1967).

In man the average cerebral blood flow is about 50–60 ml/100 g per minute (brain weight $= 1400$ g). Knowing the blood flow, one can calculate the uptake or release of any substance by determining the cerebral arteriovenous concentration difference of that substance (Fick principle). Oxygen uptake of the brain is about 3.3 ml O_2/100 g per minute. Of this amount, the white matter, which is about 60% of the brain, uses about 0.3 ml oxygen, whereas the grey matter (40%) uses about 3 ml O_2 per minute (thus grey matter has a much higher O_2 consumption). The brain is very sensitive to its blood and O_2 supply; complete ischemia for about 5 seconds leads to loss of consciousness and complete ischemia for a few minutes causes irreversible functional brain damage.

The brain produces about as much CO_2 as it takes up O_2, hence the respiratory exchange ratio, or RQ $= CO_2$ produced/O_2 consumed $= 0.99$.

This means that *glucose* is the chief source of energy of the brain. Severe hypoglycemia—eg, insulin shock—produces disturbances in brain function. Glucose is said to pass by "facilitated diffusion" across the nerve cell membranes. During *prolonged* starvation, there is appreciable utilization of other substances (eg, ketone bodies). Although cerebral capillaries are permeable to amino acids, there is very little utilization of amino acids by the brain. Most cerebral cells do not require insulin to utilize glucose.

Permeability of Capillaries of the Brain

The permeability of brain capillaries and brain cells (?) has characteristic features, which have led to the concept of "blood-brain barrier." Lipid-soluble substances pass easily (eg, alcohol, ether, chloroform, N_2 gas, and so on).

Water-soluble substances are restricted. Nutrients such as glucose, amino acids, and so on are carrier-mediated (can be saturated or inhibited by competitive overloading). Metabolic "waste products" and organic ions also seem to be carrier mediated. Some nonmetabolized noncharged solutes (eg, inulin, sucrose, mannitol) cannot diffuse across the capillary membrane.

Microscopically, the borders of the endothelial cells overlap markedly with no signs of "pores" except perhaps in the vessels of the hypothalamic area. This may partly account for the functional "barrier."

Catecholamines have difficulty in passing across the endothelium and are said to accumulate within the endothelial cells of the cerebral capillaries. Moreover, the brain capillary walls contain DOPA-decarboxylase and MAO which break down such accumulated vasoactive amines. Thus, it is evident that complex mechanisms are responsible for the blood-brain barrier (BBB).

Regulation of Cerebral Blood Flow

$$\dot{Q} = \frac{P_1 - P_2}{R}$$

where P_1 is the pressure in the internal carotid and vertebral arteries; P_2

is the pressure in the internal jugular veins; and P_1 is regulated chiefly by the baroreceptor reflexes. Gravity affects both pressures equally; therefore, it does not change P_1–P_2 in the erect posture.

Cerebral vascular resistance (R) depends on the individual length, the diameter, the number of open vessels and the blood viscosity. Of these, the diameter and the number of open vessels are the most variable.

Passive changes in diameter can occur if intracranial pressure rises (eg, subdural hemorrhage, tumors, and so on). This tends to reduce the transmural pressure, causing the vessels to collapse and inducing ischemia. The powerful Cushing reflex raises the systemic arterial pressure and maintains cerebral blood flow up to a CSF pressure of about 50 cm H_2O (body horizontal). Above this pressure, the mechanism for maintaining cerebral blood flow may fail and the subject may become comatose.

Active changes in vascular diameter may be caused by neural, chemical and mechanical influences.

Neural Control. Postganglionic sympathetic fibers are derived from the cervical ganglia and travel along the cerebral arteries. They have been traced by various methods to the cerebral arterioles up to the point where the continuous layer of smooth muscle disappears. Cutting these nerves or blocking the ganglia with drugs does not alter cerebral blood flow (hence, they do not have "tonic" influence). Electrical stimulation of sympathetic nerves has given controversial results. Early studies showed only slight reduction of cerebral flow. More recent work has demonstrated marked vasoconstriction and decreased blood flow in dogs (effect could be abolished by α-receptor blockade with drugs). Equally controversial is the effect of intracarotid injection of catecholamines on cerebral blood flow. Early investigators reported little or no effect, whereas some recent studies report a significant vasoconstrictor action.

The parasympathetic nerve fibers to the cerebral vessels are found in the facial (VII) nerve and travel by way of the geniculate ganglion and the great superficial petrosal nerve to reach the nerve plexus around the internal carotid and the Circle of Willis. Stimulation of the 7th nerve tends to produce a detectable increase in flow which is not very marked.

It may be concluded that at present the functional significance of the autonomic nerves in the control of cerebral circulation in health and disease is unclear.

Chemical Control. Much more important is the *chemical* regulation of cerebral vessels. Most powerful is the influence of CO_2 and $[H^+]$ on cerebral vascular smooth muscle. A rise in P_{CO_2} around the vascular muscle cells is a potent dilator and vice versa. Some believe that CO_2 acts by changing $[H^+]$ in interstitial fluid of the brain around vascular smooth muscle. Inhalation of 7% CO_2 doubles the cerebral blood flow if P_1-P_2 is held constant ($P_{aCO_2} = 80$ torr). Decreased arterial P_{CO_2}—from the normal 40 torr—by hyperventilation causes cerebral vasoconstriction and decreased flow which may cause dizziness and other cerebral dysfunctions. Also, blood pressure falls, respiratory alkalosis occurs, and so forth. Cerebral blood flow may fall to about 60% of normal (fig. 184).

Cerebral vessels also respond to changes in P_{O2} but to a lesser degree. Decreased arterial P_{O2} causes vasodilation and vice versa. These responses adjust cerebral blood flow to the metabolic activity of the brain. Intense neuronal activity that occurs in convulsions may increase cerebral blood flow by about 50%. Flashes of light thrown into the eyes cause increased blood flow in the occipital cortex. Also breath holding increases flow (raises the arterial P_{CO_2} and decreases the P_{O2}) by causing cerebral vasodilation (not Valsalva's maneuver). Notable is the peculiar vascular reaction to high P_{O2} seen in premature babies. Prolonged exposure to high oxygen in these babies tends to cause marked growth of the retinal vessels, which may lead to retinal detachment and blindness

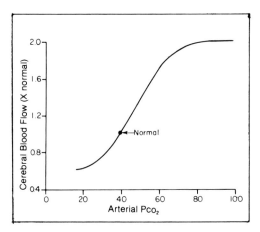

Fig. 184. The effect of arterial P_{CO_2} on cerebral vascular resistance and blood flow (perfusion pressure held constant). [Reproduced, with permission, from Guyton, A.C.: Textbook of Medical Physiology; 5th ed. (W.B. Saunders, Philadelphia 1976).]

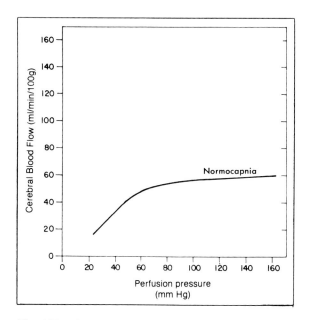

Fig. 185. Diagram showing autoregulation in the cerebral circulation (arterial P_{CO_2} held constant at normal levels). [Modified from Berne, R.M.; Levy, M.N.: Cardiovascular Physiology; 3rd ed. (C.V. Mosby, St. Louis 1977).]

(called retrolental fibroplasia). It can also occur in normal babies if oxygen is given for long periods (many days).

Brain circulation exhibits distinct autoregulation of flow when perfusion pressure is altered. In animal experiments flow remains constant when $P_1 - P_2$ is varied between 60 and 160 mmHg. Below 60 mmHg flow declines with fall of perfusion pressure (fig. 185). The mechanism of autoregulation seems to be both *metabolic* and *myogenic*.

Although blood flow in *localized* regions of the brain varies with the metabolic activity of those regions, measurement of overall cerebral blood flow with the nitrous oxide method does not reveal any change during intellectual activity, sleep, apprehension, etc. This is probably due to the fact that the change in flow is *slight* and is not detectable with the not too accurate N_2O method. Ingvar could detect regional vasodilator responses to sensory, emotional and intellectual stimuli in certain parts of the cortex with the use of [85]Kr and external recording of γ-scintillation in man. Lassen [1974] used ^{133}Xe with similar results (fig. 186).

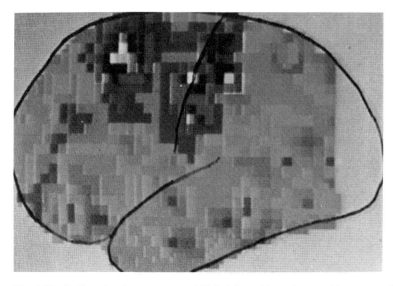

Fig. 186. Radioactive isotope xenon-133 is injected into the carotid artery and the arrival and washout of the isotope is detected with an array of scintillation counters placed on the side of the head during the 2 min after injection. The data are processed by computers and displayed on color TV. In this black and white reproduction the dark areas represent increased blood flow. The subject moved his fingers on the right side, while the left side hemisphere was scanned. Note that the active regions are the hand-finger area of the central cortex and the supplementary motor area. [Reproduced, with permission, from Lassen, N.A.; Ingvar, D.H.; Skinhoj, E.: Brain function and blood flow. Scientific American 239/4:62 (1978).]

Cerebrospinal Fluid (CSF)

The cerebrospinal fluid fills the ventricles of the brain and surrounds the brain and spinal cord in the subarachnoid space. The specific gravity of the brain and spinal cord is slightly greater than that of the CSF (isolated tissue would sink in CSF). Thus, although brain weight is about 1400 g in air, it weighs only 50 g in the CSF.

The main function of the CSF is to *cushion* the delicate structures of the brain and spinal cord from mechanical injury. Sudden movements or blows on the head do not damage, except in extreme cases such as boxing.

About 50% of the CSF is derived from the choroid plexuses and the ependymal lining of the ventricles and the remainder by filtration from the exchange vessels of the pia mater on the surface of the brain and spinal cord. The fluid from the lateral ventricles and from the third ventricle, passes down the aqueduct of Sylvius to the 4th ventricle. Then it passes out by way of the foramen of Magendie and the two foramina of Luschka into the cysterna magna and the subarachnoid spaces surrounding the spinal cord and the brain.

Most of the CSF is absorbed into the venous sinuses of the brain through the *arachnoidal villi* or *granulations,* which are very permeable to all constituents—including the small amount of protein—of CSF. The volume absorbed is estimated to be from 500 to 1000 ml per day in man. Only a small amount of CSF reenters the capillaries of the pia mater of the brain and spinal cord.

There are important differences in the composition of CSF and plasma. The most striking is the *very low protein concentration* of CSF (20 mg vs. 7000 mg per 100 ml fluid), *low glucose* (60 mg vs. 100 mg/dl, *low K^+* (2.9 mEq vs. 4.6 mEq per 100 ml), and *low pH* (7.3 vs. 7.4). The CSF is considered to be the result of *filtration across* the capillaries and the *secretion* of epithelial cells of the choroid plexuses. There are no leukocytes or RBC in CSF.

The volume of CSF is about 150 ml, depending on the relation between its rate of *formation* and rate of *absorption.* Increased formation or decreased absorption will increase the volume of fluid and the pressure of CSF. Increased CSF pressure tends to obstruct venous outflow from the retina and cause edema of the optic disc (papilledema), an important diagnostic sign. Subdural hemorrhage or meningitis causes a large number of blood cells to enter CSF; these can block the arachnoid villi and raise the pressure.

Newborn babies may have disturbances of the production and circulation of CSF. In one form there is overdevelopment of the choroid plexus which secretes at such a rate that the arachnoid granulations cannot drain. This results in excess fluid inside and outside of the brain, causing a large head and pressure atrophy of the brain. This type is known as *communicating* or *hypersecretory hydrocephalus.* Another type is due to stenosis of the aqueduct of Sylvius, causing enlargement of the lateral and the third ventricles. This is known as *obstructive or noncommunicating hydrocephalus.*

Compression of *both* internal jugular veins raises the CSF pressure in the cranial and lumbar regions by increasing the volume of blood in the vessels of the cranial cavity. If the spinal canal is completely blocked by a tumor, the pressure recorded in the lumbar region will not change when the jugulars are compressed. This maneuver is called the Queck-enstedt test. If the lumbar pressure rises, the test is said to be positive, which is normal; if it does not rise, the test is negative and indicates a spinal block.

Reference

Lassen, N.A.: Control of cerebral circulation in health and disease. Circ. Res. *34*:749–760 (1974).

Appendix

A1 Review of Some Physical Units

Metric Units

cgs (centimeter-gram-second) system, or
mks (meter-kilogram-second) system (now universally adopted)
Acceleration = change in velocity
Unit of acceleration: $a = 1$ cm/sec^2 or 1 m/sec^2
Acceleration of gravity: g = about 980 cm/sec^2 or 9.8 m/sec^2 at sea level (g does not vary more than 0.3% depending on local altitude and an approximate value of 1000 cm/sec^2 or 10 m/sec^2 may be handy for a quick computation or for a quick check of more precise computation)
Unit of Mass = gram or kilogram (physiologists still use the gram as a unit of force).
Force = M · a

1 g mass accelerated by 1 cm/sec$^2 = \dfrac{1 \text{ g} \cdot \text{cm}}{\text{sec}^2} = 1$ dyne

1 kg mass accelerated by 1 m/sec$^2 = \dfrac{1 \text{ kg} \cdot \text{m}}{\text{sec}^2}$

$\qquad\qquad\qquad\qquad = 1$ newton $(= 10^5$ dynes)

Force of gravity acting on 1 g mass = 980 dynes
Force of gravity acting on 1 kg mass = 9.8 newtons
Pressure = Force/unit area = 1 dyne/cm^2 or 1 newton/m^2 = $\boxed{\text{pascal}}$
$\qquad\qquad\qquad\qquad$ (kPa = 1000 pascals)
(The pascal is a small unit of pressure.)
Work (or Energy) = Force · distance = F · s = M · a · s

1 dyne force $\left(\dfrac{\text{g} \cdot \text{cm}}{\text{sec}^2}\right)$ acting for a distance of 1 cm =

$\qquad\qquad\qquad$ 1 erg or dyne-cm $= \left(\dfrac{\text{g} \cdot \text{cm}^2}{\text{sec}^2}\right)$

1 newton force $\left(\dfrac{\text{kg} \cdot \text{m}}{\text{sec}^2}\right)$ acting for a distance of 1 m $= 1\dfrac{\text{kg} \cdot \text{m}^2}{\text{sec}^2} =$

$\boxed{\textit{1 joule} \text{ or } \textit{newton-meter}}$ (1 joule $= 10^7$ ergs)

Force acting against gravity on 1 kg mass for a distance of 1 m does 9.8 joules of work.

Power = work/unit time, 1 joule/sec = $\boxed{\text{1 watt}}$

Pressure of a column of liquid = height of column × density of liquid × acceleration of gravity

Question: A column of mercury is often used to measure pressure (density = 13.6 g/cm³ or 1.36 × 10⁴ kg/m³). What would be the pressure (cgs) of a column of Hg that is 11 cm high?

Answer: 11 cm · 13.6 g/cm³ · 980 cm/sec² = 146 608 dynes/cm².

Question: Calculate the pressure in pascals.

Answer: .11 m × 1.36 × 10⁴ kg/m³ × 9.8 m/sec² = 14,660.8 Pa.

A2 Auscultatory Method of Determining Arterial Pressure in Man

Principle

Normal flow of blood in the peripheral arteries is laminar (or streamline) and no sounds are heard if a stethoscope is applied over any large artery. However, if a large artery in an extremity is completely occluded by the external pressure of an inflated cuff that surrounds the extremity and then the pressure is gradually reduced, the escape of blood under the cuff with each heart beat will cause *turbulent flow* in the artery distal to the occlusion and *vibrations* of the vascular wall. Such vibrations will cause sounds (known as Korotkoff sounds) that can be heard with a stethoscope over the vessel. When the flow in the distal artery becomes sufficiently non-turbulent the sounds disappear. This occurs when cuff pressure is somewhat below the diastolic pressure in the compressed artery.

Apparatus

Stethoscope (diaphragm or bell type); Sphygmomanometer (mercury or aneroid type).

Method

The artery that is conveniently used is the brachial, either the right or the left. Clothing must be removed to avoid constriction of the vessel and permit the application of the arm bag (cuff). The subject may be seated or lying down, but in either case the *brachial artery must be positioned at the level of the heart* to avoid changes in pressure due to the weight of the column of blood in the arterial system. The observer should be *seated* comfortably. Time should be allowed for recovery from recent exercise or excitement.

The standard adult-sized cuff (12–13 cm wide and 23 cm long) is applied snugly around the arm with the lower edge about an inch above the antecubital space. If a mercury manometer is used, the apparatus must be on a level surface and the level of mercury should be at the zero mark. The cuff is rapidly inflated to a pressure that abolishes the radial pulse (this occurs when cuff pressure exceeds the systolic pressure in the brachial artery). There should be no undue bulging of the bag. The diaphragm or bell of the stethoscope is placed below the cuff *over the artery* (medial to biceps tendon) at the bend of the elbow avoiding contact with the cuff. Enough pressure is used to insure contact everywhere around the rim, but not hard enough to deform or occlude the artery. The cuff is deflated by opening the valve permitting the pressure to drop at a slow rate (2–3 mmHg per sec). As the pressure is lowered a series of sounds (Korotkoff) are heard with each heart beat.

At first it is a faint sound with each heart beat, the *appearance* of which is taken to represent the *systolic* (or peak) pressure in the artery.

Then the sounds become louder and more distinct.

As the cuff pressure is lowered further, at a certain point the sounds suddenly become dull or *muffled*. This phase is usually found to correspond most closely to the *diastolic* arterial pressure (some investigators do not agree with this view).

Finally, about 10 mmHg below the muffled phase, the sounds disappear completely. At this point the flow of blood in the artery under the stethoscope is nonturbulent during the entire cardiac cycle, hence no sounds are heard. The physical basis for the changes in the character of the sounds with the different phases is complex and not well understood.

The bag is *completely* deflated to relieve the congestion and the discomfort in the forearm and the procedure repeated to determine the systolic and diastolic pressures *several times*. The values obtained after several determinations are said to be closer to the normal values.

Occasionally, one may note in individuals with advanced arteriosclerosis that the sounds are never lost even though the cuff is completely deflated. This means that the artery under the stethoscope is so diseased that the flow is turbulent even without the cuff (localized narrowing causes high velocity which tends to cause turbulence).

The American Committee for Standardizing Blood Pressure Measurements advised that the beginning of both the muffled phase and the disappearance of sounds be recorded as follows: 120/80–70, but few physicians have followed this recommendation.

Sometimes the standing posture causes a marked drop of blood pressure (postural or orthostatic hypotension) and blood pressure must be taken in this position to detect the condition. It is important under these circumstances to make sure that the arm is at the heart level to avoid erroneous determinations.

Occasionally, blood pressure in the lower extremity is to be measured. This is done at the popliteal artery and a larger cuff is applied over the lower thigh in the recumbent position. Smaller cuffs for infants and children are also available.

Note. Practicing physicians should occasionally check the accuracy of their manometers (particularly the aneroid). Sometimes the mercury manometer becomes inaccurate when the porous membrane at the top of the glass tube gets clogged. Air does not enter *freely* during deflation and the readings become erroneously high (false diagnosis of hypertension!).

A3 Measurement of Coronary Flow in Intact Animals and Man

Nitrous Oxide Method

In a supine subject the coronary sinus is catheterized and a large artery punctured to secure several blood samples from both vessels at frequent intervals. A mixture of 15% N_2O, 21% O_2 and 64% N_2 is then inhaled for about 10 to 15 min. During this period frequent blood samples are withdrawn simultaneously from both vessels at known intervals after beginning inhalation. The samples are analyzed for N_2O concentration and the values plotted against the time base. At the end of the period the arterial blood and the coronary sinus blood should have about the same concentration of N_2O, indicating that the heart tissue has been saturated with nitrous oxide (fig. 187).

Calculation of coronary flow is based on a special application of the Fick principle as follows:

$$\text{Coronary flow (per unit time per unit mass)} = \frac{\text{Total uptake of } N_2O \text{ by LV per unit mass of tissue during saturation period}}{\text{Total arteriovenous } N_2O \text{ concentration difference during saturation period}}$$

The total amount of N_2O taken up by a unit mass (100 g) of left ventricle is found as follows: at the end of 10 to 15 min equilibrium is reached between blood and heart tissue with respect to N_2O. At this time, knowing the N_2O content of coronary venous blood can provide the amount of N_2O in heart muscle if the partition coefficient for N_2O between blood and heart is known.

$$\text{Partition coefficient for } N_2O \text{ in heart tissue} = \frac{\text{Amount of } N_2O/g \text{ heart at } 37°\text{ C}}{\text{Amount of } N_2O/ml \text{ venous blood at } 37°\text{ C}}$$

Experiments in animals have shown that the coefficient is close to 1.0. Therefore, 100 ml coronary sinus blood at the end of 15 minutes will have the same amount of N_2O as 100 g of left ventricular tissue (sinus drains mainly the left ventricle). Thus, the equation may be written as follows:

$$\text{Left ventricular coronary flow (ml/min per 100 g)} = \frac{\text{Amount of } N_2O \text{ in 100 ml coronary sinus blood at } \textit{end} \text{ of 15 min}}{\int_0^{15} (aN_2O - vN_2O)\, dt}$$

aN_2O = arterial N_2O concentration in ml/dl

vN_2O = venous N_2O concentration in ml/dl

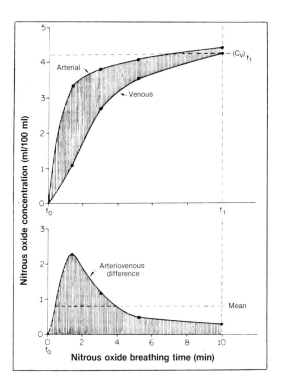

Fig. 187. Changes in arterial and venous concentration of N_2O during the inhalation of 15% N_2O with 21% oxygen and 64% nitrogen. The mean arteriovenous concentration difference of N_2O (*lower curve*) is used to calculate coronary blood flow. [Reproduced, with permission, from D'Alecy, L.G.: The cerebral circulation; in Ruch and Patton: Physiology and Biophysics; 20th ed.; vol. II: Circulation, Respiration and Fluid Balance. (W.B. Saunders, Philadelphia 1974).]

Another way of putting the equation is:

$$\text{LV coronary flow (ml/min per 100 g)} = \frac{\begin{array}{c}N_2O \text{ absorbed (average/min)} \\ \text{by } 100 \text{ g heart}\end{array}}{\text{Mean } (aN_2O - vN_2O) \text{ per min}}$$

Indicator Washout or Clearance Method

A radioactive material (eg, [86]rubidium, [85]krypton, [133]xenon) is injected into the ostium of one of the coronary arteries through a catheter and the activity on the surface of the chest over the precordium is counted continuously. As the indicator diffuses rapidly into the myocardium and then is washed away from the heart, the activity over the precordium declines in a multi-exponential manner with respect to time. A semilogarithmic plot will show an *initial* straight line, the slope

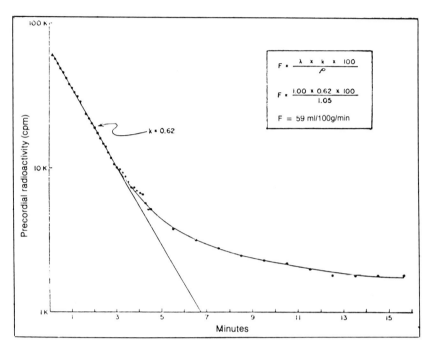

Fig. 188. Myocardial washout method for determining coronary flow. At time zero, 100 μCi of ^{85}Kr is injected into the left coronary artery. The precordial radioactivity (counts/min) is recorded for a period of time. A semilog plot is drawn and the initial slope of the decline indicates the rate of removal of ^{85}Kr by coronary blood (coronary flow rate). [Reproduced, with permission, from Herd, J.A.; Hollenberg, M.; Thorburn, G.D.; Kopald, H.H.; Barger, A.C.: Myocardial blood flow determined with krypton-85 in unanesthetized dogs. Am. J. Physiol. *203*:122 (1962).]

of which represents the coronary flow rate (fig. 188). The higher the flow rate, the greater will be the initial slope of the line.

Coronary flow   = slope × partition coefficient of indicator × 100 g tissue.
(ml/min per 100 g)   (k)   (λ)

This method has the advantage of allowing the measurement of flow in the right coronary circulation in man.

Measurement of Coronary Flow Velocity

A catheter-tip velocity transducer (electromagnetic) is placed in the orifice of a coronary artery or in the coronary sinus. The instrument measures velocity of flow and not volume flow as in the previous two methods.

Velocity of flow $= \dfrac{\dot{Q}}{\text{x-area}}$.

Index